SEWAGE AND LANDFILL LEACHATE

Assessment and Remediation of Environmental Hazards

SEWAGE AND LANDFILL LEACHATE
Assessment and Remediation of Environmental Hazards

Edited by
Marco Ragazzi

Apple Academic Press Inc. | Apple Academic Press Inc.
3333 Mistwell Crescent | 9 Spinnaker Way
Oakville, ON L6L 0A2 | Waretown, NJ 08758
Canada | USA

©2016 by Apple Academic Press, Inc.

Exclusive worldwide distribution by CRC Press, a member of Taylor & Francis Group

No claim to original U.S. Government works

Printed in the United States of America on acid-free paper

International Standard Book Number-13: 978-1-77188-394-8 (Hardcover)

International Standard Book Number-13: 978-1-77188-395-5 (eBook)

This book contains information obtained from authentic and highly regarded sources. Reprinted material is quoted with permission and sources are indicated. Copyright for individual articles remains with the authors as indicated. A wide variety of references are listed. Reasonable efforts have been made to publish reliable data and information, but the authors, editors, and the publisher cannot assume responsibility for the validity of all materials or the consequences of their use. The authors, editors, and the publisher have attempted to trace the copyright holders of all material reproduced in this publication and apologize to copyright holders if permission to publish in this form has not been obtained. If any copyright material has not been acknowledged, please write and let us know so we may rectify in any future reprint.

Trademark Notice: Registered trademark of products or corporate names are used only for explanation and identification without intent to infringe.

Library and Archives Canada Cataloguing in Publication

Sewage and landfill leachate : assessment and remediation of environmental hazards / edited by Marco Ragazzi.

Includes bibliographical references and index.
Issued in print and electronic formats.
ISBN 978-1-77188-394-8 (hardcover).--ISBN 978-1-77188-395-5 (pdf)
1. Sewage--Purification. 2. Sewage--Management. 3. Sanitary landfills--Leaching. 4. Sewage sludge--Environmental aspects. 5. Leachate--Environmental aspects. I. Ragazzi, Marco, author, editor

TD745.S49 2016   628.3   C2016-901764-8   C2016-901765-6

Library of Congress Cataloging-in-Publication Data

Names: Ragazzi, Marco, editor.
Title: Sewage and landfill leachate : assessment and remediation of environmental hazards / editor, Marco Ragazzi.
Description: New Jersey : Apple Academic Press Toronto, [2016] | Includes bibliographical references and index.
Identifiers: LCCN 2016011414| ISBN 9781771883948 (hardcover : acid-free paper) | ISBN 9781771883955 (eBook)
Subjects: LCSH: Sanitary landfills--Leaching. | Sewage--Purification.
Classification: LCC TD795.7 .S49 2016 | DDC 628.3/9--dc23
LC record available at http://lccn.loc.gov/2016011414

Apple Academic Press also publishes its books in a variety of electronic formats. Some content that appears in print may not be available in electronic format. For information about Apple Academic Press products, visit our website at **www.appleacademicpress.com** and the CRC Press website at **www.crcpress.com**

ABOUT THE EDITOR

MARCO RAGAZZI

Marco Ragazzi has a PhD in sanitary engineering from Milan Polytechnic, Italy. The author or co-author of more than 500 publications (111 in the Scopus database), he is currently a member of the Department of Civil, Environmental, and Mechanical Engineering at the University of Trento, Italy. His research interests include solid waste and wastewater management, environmental engineering, and environmental impact risk assessment.

CONTENTS

Acknowledgement and How to Cite .... *ix*

List of Contributors .... *xi*

Introduction .... *xv*

Part I: Overview of Sewage and Leachate Treatments

1. Sustainable Treatment of Landfill Leachate 3

Mohamad Anuar Kamaruddin, Mohd. Suffian Yusoff, Hamidi Abdul Aziz, and Yung-Tse Hung

2. Sustainability of Domestic Sewage Sludge Disposal 31

Claudia Bruna Rizzardini and Daniele Goi

Part II: Sewage Sludge Treatments

3. Composting Used as a Low Cost Method for Pathogen Elimination in Sewage Sludge in Mérida, Mexico 49

Dulce Diana Cabañas-Vargas, Emilio. de los Ríos Ibarra, Juan. P. Mena-Salas, Diana Y. Escalante-Réndiz, and Rafael Rojas-Herrera

4. An Experimental Investigation of Sewage Sludge Gasification in a Fluidized Bed Reactor 63

L. F. Calvo, A. I. García, and M. Otero

5. Bacterial Consortium and Axenic Cultures Isolated from Activated Sewage Sludge for Biodegradation of Imidazolium-Based Ionic Liquid 83

M. Markiewicz, J. Henke, A. Brillowska-Dąbrowska, S. Stolte, J. Łuczak, and C. Jungnickel

Part III: Assessment of Sewage Sludge Hazard, Pre- and Post-Treatment

6. Seeking Potential Anomalous Levels of Exposure to PCDD/Fs and PCBs through Sewage Sludge Characterization 103

E. C. Rada, M. Schiavon, and M. Ragazzi

7. Occurrence and Distribution of Synthetic Organic Substances in Boreal Coniferous Forest Soils Fertilized with Hygienized Municipal Sewage Sludge 121

Richard Lindberg, Kenneth Sahlén, and Mats Tysklind

viii   Contents

Part IV: Leachate Treatments

8. Analysis of Electro-Oxidation Suitability for Landfill Leachate Treatment through an Experimental Study 149

Elena Cristina Rada, Irina Aura Istrate, Marco Ragazzi, Gianni Andreottola, and Vincenzo Torretta

9. Potential of Ceria-Based Catalysts for the Oxidation of Landfill Leachate by Heterogeneous Fenton Process 173

E. Aneggi, V. Cabbai, A. Trovarelli, and D. Goi

Part V: Assessment of Leachate Hazards, Pre- and Post-Treatment

10. Investigation of Physicochemical Characteristics and Heavy Metal Distribution Profile in Groundwater System Around the Open Dump Site 197

S. Kanmani and R. Gandhimathi

11. Application of Response Surface Methodology (RSM) for Optimization of Semi-Aerobic Landfill Leachate Treatment Using Ozone 225

Salem S. Abu Amr, Hamidi Abdul Aziz, and Mohammed J. K. Bashir

12. Removal of COD, Ammoniacal Nitrogen and Colour from Stabilized Landfill Leachate by Anaerobic Organism 247

Mohamad Anuar Kamaruddin, Mohd Suffian Yusoff, Hamidi Abdul Aziz, and Nur Khairiyah Basri

Author Notes 269

Index 273

ACKNOWLEDGMENT AND HOW TO CITE

The editor and publisher thank each of the authors who contributed to this book. The chapters in this book were previously published elsewhere. To cite the work contained in this book and to view the individual permissions, please refer to the citation at the beginning of each chapter. Each chapter was read individually and carefully selected by the editor; the result is a book that provides a multiperspective look at research into many elements of remediating environmental hazards connected to sewage and landfill leachate. The chapters included examine the following topics:

Part I: Overview of Sewage and Leachate Treatments
- Chapter 1 summarizes leachate sustainable treatment processes, including biological, physical, and chemical techniques, reported from 2008 to 2012.
- Chapter 2 evaluates the level of pollution of sewage sludge and amended agricultural soils, applying the dictates of the Third Working document on sludge.

Part II: Sewage Sludge Treatments
- Chapter 3 tests the compost process as an economical and efficient method for pathogen elimination in sludge from a municipal wastewater system.
- Chapter 4 identifies characteristics of sewage sludge gasification in an atmospheric fluidized-bed gasifier.
- Chapter 5 attempts to identify the exact microbial strains which might be partaking in the process of biological degradation ionic liquids, in order to uncover species especially predisposed to degrade ionic liquids.

Part III: Assessment of Sewage Sludge Hazard, Pre- and Post-Treatment

- Chapter 6 proposes a methodology to detect food-chain anomalies as a consequence of the release of persistent organic pollutants in the air and their subsequent deposition to farmlands and pastures.
- Chapter 7 evaluates the occurrence and distribution of selected synthetic organic substances in relevant matrices following application of dried and granulated municipal sewage sludge to boreal coniferous forest soils; the authors then perform an initial environmental risk assessment.

Part IV: Leachate Treatments

- Chapter 8 seeks to contribute to the knowledge of electrochemical treatments for the reduction of chemical oxygen demand (COD), biochemical oxygen demand (BOD5), ammonium, and total suspended solids in leachate; it also observes whether there was any hexavalent chromium in the liquid sample due to the different oxidative conditions of this treatment.
- Chapter 9 investigates doped ceria materials in the treatment of landfill leachate by a heterogeneous Fenton process.
- Chapter 10 aims to develop an understanding of the natural groundwater quality in an Tamil Nadu open dumping site and the adjacent areas through the dug wells and bore wells that were selected for this purpose.
- In Chapter 11, the statistical relationships among three independent factors (ozone dosage, COD concentration, and reaction time) for the treatment of semi-aerobic stabilized leachate were assessed through response surface methodology (RSM).
- Chapter 12 investigates the effect of leachate parameters removal by anaerobic organism cultures.

LIST OF CONTRIBUTORS

Hamidi Abdul Aziz
School of Civil Engineering, Universiti Sains Malaysia, 14300 Nibong Tebal, Pulau Penang, Malaysia

Salem S. Abu Amr
School of Civil Engineering, Engineering Campus, Universiti Sains Malaysia, 14300, Nibong Tebal, Penang, Malaysia

Gianni Andreottola
Department of Civil, Environmental and Mechanical Engineering, University of Trento, via Mesiano 77, Trento I-38123, Italy

E. Aneggi
Department of Chemistry, Physics and Environment, University of Udine, Via del Cotonificio, 108-331100 Udine, Italy

Mohammed J. K. Bashir
Department of Environmental Engineering, Faculty of Engineering and Green Technology, Universiti-Tunku Abdul Rahman, 31900 Kampar, Perak, Malaysia

Nur Khairiyah Basri
School of Civil Engineering, Universiti Sains Malaysia, 14300 Nibong Tebal, Pulau Penang, Malaysia

A. Brillowska-Dąbrowska
Department of Microbiology, Chemical Faculty, Gdańsk University of Technology, ul. Narutowicza 11/12, 80-233 Gdańsk, Poland

Dulce Diana Cabañas-Vargas
Faculty of Chemical Engineering, Autonomous University of Yucatán, Campus de Ciencias Exactas e Ingenierias, Periférico Nte. Km 33.5 Col. Chuburná de Hidalgo Inn, Mérida Yucatán, 97203, Mexico

V. Cabbai
Department of Chemistry, Physics and Environment, University of Udine, Via del Cotonificio, 108-331100 Udine, Italy

L. F. Calvo
IMARENABIO, University of León, Avenida de Portugal 41, 24071 León, Spain

Diana Y. Escalante-Réndiz
Faculty of Chemical Engineering, Autonomous University of Yucatán, Campus de Ciencias Exactas e Ingenierias, Periférico Nte. Km 33.5 Col. Chuburná de Hidalgo Inn, Mérida Yucatán, 97203, Mexico

R. Gandhimathi
Department of Civil Engineering, National Institute of Technology, Tiruchirappalli, Tamil Nadu, India

A. I. García
IMARENABIO, University of León, Avenida de Portugal 41, 24071 León, Spain

Daniele Goi
Chemistry Physics Environment Department, University of Udine, via del Cotonificio 108, 33100 Udine, Italy

J. Henke
Department of Chemical Technology, Chemical Faculty, Gdańsk University of Technology, ul. Narutowicza 11/12, 80-233 Gdańsk, Poland

Yung-Tse Hung
Department of Civil and Environmental Engineering, Cleveland State University, Cleveland, OH 44115, USA

Emilio. de los Ríos Ibarra
Independent Researcher. Campus de Ciencias Exactas e Ingenierias, Periférico Nte, Km 33.5 Col. Chuburná de Hidalgo Inn, Mérida Yucatán 97203, Mexico

Irina Aura Istrate
Department of Civil, Environmental and Mechanical Engineering, University of Trento, via Mesiano 77, Trento I-38123, Italy; Department of Energy Production and Use, University Politehnica of Bucharest, Splaiul Independentei 313, Bucharest 060042, Romania

C. Jungnickel
Department of Chemical Technology, Chemical Faculty, Gdańsk University of Technology, ul. Narutowicza 11/12, 80-233 Gdańsk, Poland

Mohamad Anuar Kamaruddin
School of Civil Engineering, Universiti Sains Malaysia, Engineering Campus, 14300 Nibong Tebal, Penang, Malaysia

S. Kanmani
Department of Civil Engineering, National Institute of Technology, Tiruchirappalli, Tamil Nadu, India

Richard Lindberg
Department of Chemistry, Umeå University, Umeå 90187, Sweden

J. Łuczak
Department of Chemical Technology, Chemical Faculty, Gdańsk University of Technology, ul. Narutowicza 11/12, 80-233 Gdańsk, Poland

M. Markiewicz
Department of Chemical Technology, Chemical Faculty, Gdańsk University of Technology, ul. Narutowicza 11/12, 80-233 Gdańsk, Poland

Juan. P. Mena-Salas
Faculty of Chemical Engineering, Autonomous University of Yucatán, Campus de Ciencias Exactas e Ingenierias, Periférico Nte. Km 33.5 Col. Chuburná de Hidalgo Inn, Mérida Yucatán, 97203, Mexico

List of Contributors

M. Otero
CESAM, Department of Chemistry, University of Aveiro, Campus de Santiago, 3810-193 Aveiro, Portugal; Department of Applied Chemistry and Physics, University of León, Campus de Vegazana, 24071 León, Spain

Elena Cristina Rada
Department of Civil, Environmental and Mechanical Engineering, University of Trento, via Mesiano 77, Trento I-38123, Italy

Marco Ragazzi
Department of Civil, Environmental and Mechanical Engineering, University of Trento, via Mesiano 77, Trento I-38123, Italy

Claudia Bruna Rizzardini
Chemistry Physics Environment Department, University of Udine, via del Cotonificio 108, 33100 Udine, Italy

Rafael Rojas-Herrera
Faculty of Chemical Engineering, Autonomous University of Yucatán, Campus de Ciencias Exactas e Ingenierias, Periférico Nte. Km 33.5 Col. Chuburná de Hidalgo Inn, Mérida Yucatán, 97203, Mexico

Kenneth Sahlén
Department of Forest Ecology and Management, Swedish University of Agricultural Sciences, Umeå 90183, Sweden

M. Schiavon
Department of Civil Environmental and Mechanical Engineering, Via Mesiano, University of Trento, Italy; Fondazione Trentina per la Ricerca sui Tumori, Corso III Novembre 162, I-38122–Trento, Italy

S. Stolte
Center for Environmental Research and Sustainable Technology, University of Bremen, UFT Leobener Strasse, 28359 Bremen, Germany

Vincenzo Torretta
Department of Science and High Technology, University of Insubria, Via G.B. Vico 46, Varese I-21100, Italy

A. Trovarelli
Catalysis Group, Department of Chemistry, Physics and Environment, University of Udine, Via del Cotonificio, 108-331100 Udine, Italy

Mats Tysklind
Department of Chemistry, Umeå University, Umeå 90187, Sweden

Mohd. Suffian Yusoff
School of Civil Engineering, Sains Malaysia, 14300 Nibong Tebal, Pulau Penang, Malaysia

INTRODUCTION

Sewage and landfill leachate treatments include various processes that are used to manage and dispose of the liquid portions of solid waste. Untreated leachate and sewage are hazards to the environment if they enter the water system. The goal of treatment is to reduce the contaminating load to the point that leachate and sewage liquids may be safely released into groundwater, streams, lakes, and the ocean.

Around the world, however, huge volumes of contaminated water from sewage and landfill leachate are still pumped directly into water systems, especially in the world's developing nations. Aside from the damage to marine environments and fisheries that this causes, it also jeopardizes the world's vulnerable water resources.

Effective sewage management is essential for nutrient recycling and for maintaining ecosystem integrity. It is also important for improving the environment through proper drainage and disposal of wastewater; preventing floods through removal of rainwater; and preserving receiving water quality. Treatment processes facilitate the achievement of water-quality objectives.

A range of technologies are available for the treatment of sewage and landfill leachate. These include biological treatments, reverse osmosis, and chemical-physical processes. The articles included in this compendium offer important research that will help us both assess our existing treatment facilities, as well as build better, more effective ones for the future.

—Marco Ragazzi

Landfill leachate is a complex liquid that contains excessive concentrations of biodegradable and non-biodegradable products, including organic matter, phenols, ammonia nitrogen, phosphate, heavy metals, and sulfide. If not properly treated and safely disposed, landfill leachate could be an impending source of surface and ground water contamination, since it may percolate throughout soils and subsoils, causing adverse impacts to receiving waters. Lately, various types of treatment methods have been proposed to alleviate the risks of untreated leachate. However, some of the available techniques remain complicated, expensive and generally require definite adaptation during process. In chapter 1, a review of literature reported from 2008 to 2012 on sustainable landfill leachate treatment technologies is discussed, which includes biological and physical–chemical techniques.

Activated sludge is now one of the most widely used biological processes for the treatment of wastewaters from medium to large populations. It produces high amounts of sewage sludge that can be managed and perceived in two main ways: as a waste, it is discharged in landfill, as a fertilizer, it is disposed in agriculture with direct application to soil or subjected to anaerobic digestion and composting. Other solutions, such as incineration or production of concrete, bricks, and asphalt, play a secondary role in terms of their degree of diffusion. The agronomical value of domestic sewage sludge may be hidden by the presence of several pollutants such as heavy metals, organic compounds, and pathogens. In this way, the sustainability of sewage sludge agricultural disposal requires a value judgment based on knowledge and evaluation of the level of pollution of both sewage sludge and soil. Chapter 2 analyzes a typical Italian case study, a water management system of small communities, applying the criteria of evaluation of the last official document of European Union about sewage sludge land application, the "Working Document on Sludge (3rd draft, 2000)." The report brought out good sewage sludge from small wastewater treatment plants, suggesting a sustainable application.

Spreading sewage sludge from municipal wastewater treatment on land is still a common practice in developing countries. However, it is well known that sewage sludge without special treatment contains various pollutants, which are (re)introduced into the environment by sludge landspreading, and which might in turn have harmful effects on the environment and human health. This is more dangerous in places like Merida, Mexico, where soil is calcareous, with fractures along the ground and thin layers of humus. Consequently, any liquid and semisolid wastes have the potential of percolating to the subsurface and contaminating the aquifer. Chapter 3 uses composting as a low-cost process to eliminate pathogens

Introduction xvii

contained in sewage sludge from municipal wastewater treatment in order to use the final product for land spreading in a safe way for both environment and human health. Two piles for composting process at real scale were settled using a mixture of sewage sludge from municipal wastewater and green waste. Composting was carried out by windrow process, and it was monitored during four weeks. Concentrations of helminth eggs, salmonella, and fecal coliforms were measured twice a week to observe its behavior; as a control process, temperature, moisture content, and pH were also measured. After 30 days of composting sludge from municipal waste water system, salmonella was eliminated by 99%, fecal coliforms by 96%, and helminth eggs by 81%. After 3 months, compost reached GI = 160%, indicating no phytotoxicity to seeds.

In chapter 4, the gasification of sewage sludge was carried out in a simple atmospheric fluidized bed gasifier. Flow and fuel feed rate were adjusted for experimentally obtaining an air mass:fuel mass ratio (A/F) of $0.2 < A/F < 0.4$. Fuel characterization, mass and power balances, produced gas composition, gas phase alkali and ammonia, tar concentration, agglomeration tendencies, and gas efficiencies were assessed. Although accumulation of material inside the reactor was a main problem, this was avoided by removing and adding bed media along gasification. This allowed improvement of the process heat transfer and, therefore, gasification efficiency.

Extensive research and increasing number of potential industrial applications made ionic liquids (ILs) important materials in design of new, cleaner technologies. Together with the technological applicability, the environmental fate of these chemicals is being considered, and significant efforts are being made in designing strategies to mitigate their potential negative impacts. Many ILs are proven to be poorly biodegradable and relatively toxic. Bioaugmentation is known as one of the ways of enhancing the microbial capacity to degrade xenobiotics by addition of specialized strains. Chapter 5 aims to select microbial species that could be used for bioaugmentation in order to enhance biodegradation of ILs in the environment. The authors subjected activated sewage sludge to the selective pressure of 1-methyl-3-octylimidazolium chloride ([OMIM][Cl]) and isolated nine strains of bacteria that were able to prevail in these conditions. Subsequently, they utilized axenic cultures (pure cultures) of these bacteria as well as mixed consortium to degrade this IL. In addition, the authors performed growth inhibition tests and found that bacteria were able to grow in 2 mM, but not in 20 mM solutions of [OMIM][Cl]. The biodegradation conducted by the isolated consortium was higher than conducted by the activated sewage sludge when normalized by the cell density, which indicates that the isolated strains seem specifically suited to degrade the IL.

An approach to detect anomalies in the exposure to persistent organic pollutants (POPs) throughout the food chain is presented in chapter 6. The authors' proposed method is useful also for preventing soil contamination by POPs that would require a remediation intervention. A steel-making plant and its surrounding area were selected as a case study. To investigate the possible effects of the plant on the settled population, sewage sludge samples from four wastewater treatment plants were taken: one of these was chosen as reference for the population exposed to the emissions of the mill; the remaining three plants were chosen to provide background information about the POP content in sludge. No clear anomalies in dioxins (PCDD/Fs) were detected for the potentially exposed population. In terms of polychlorinated biphenyls (PCBs), the steel plant-influenced wastewater treatment plant showed a total concentration between 2.7 and 4.8 times higher than the other plants; in terms of equivalent toxicity, only slightly higher concentrations were found for the steel plant-influenced wastewater treatment plant. Therefore, if considering acceptable the daily intake from the diet of the unexposed population, the absence of a dioxin and dioxin-like emergency in the area of the mill is demonstrated. This method represents an innovative and technically simple tool to assess situations of permanent exposure to POP levels that are higher than the background

The occurrence and distribution of synthetic organic substances following application of dried and granulated (hygienized) municipal sewage sludge in Swedish boreal coniferous forests were investigated in chapter 7. Elevated concentrations of triclosan (TCS), polybrominated diphenyl ethers (PBDEs), and PCBs were detected in the humus layer. Concentrations of ethinyl estradiol (EE2), norfloxacin, ciprofloxacin, ofloxacin (FQs), and polyaromatic hydrocarbons (PAHs) were not significantly influenced. Fertilization did not alter the levels of the substances in mineral soil, ground water, and various types of samples related to air. The authors indicate that further research within this area is needed, including

Introduction

ecotoxicological effects and fate, in order to improve the knowledge regarding the use of sludge as a fertilizing agent. They call for continuous annual monitoring, with respect to sampling and analysis, on the already-fertilized fields.

Chapter 8 examines the efficiency of electro-oxidation used as the single pretreatment of landfill leachate. The experiments were performed on three different types of leachate. The results obtained using this electrochemical method results were analyzed after seven days of treatment. The main characteristics of leachate and a diagram of the experimental apparatus are presented. The overall objectives were to contribute to the knowledge of electrochemical treatments for the reduction of chemical oxygen demand (COD), biochemical oxygen demand, ammonium, and total suspended solids, and also to examine whether there was any resulting hexavalent chromium in the liquid sample. The yields obtained were considered satisfactory, particularly given the simplicity of this technology. Like all processes used to treat refluent water, the applicability of this technique to a specific industrial refluent needs to be supported by feasibility studies to estimate its effectiveness and optimize the project parameters. The authors indicate this could be a future development of the work.

In chapter 9, ceria and ceria-zirconia solid solutions were tested as catalyst for the treatment of landfill leachate with a Fenton-like process. The catalysts considered in this work were pure ceria and ceria-zirconia solid solutions as well as iron-doped samples. All the catalysts were extensively characterized and applied in batch Fenton-like reactions by a close batch system, the COD and total organic carbon (TOC) parameters were carried out before and after the treatments in order to assay oxidative abatement. Results show a measurable improvement of the TOC and COD abatement using ceria-based catalysts in Fenton-like process. The best result was achieved for iron-doped ceria-zirconia solid solution. The authors' outcomes point out that heterogeneous Fenton technique could be effectively used for the treatment of landfill leachate. They indicate that making it the object of further investigations would be worthwhile.

In chapter 10, the characterization of solid waste and the effect of the leachate from an open dumping site in Ariyamangalam, Tiruchirappalli District, Tamil Nadu, on groundwater is investigated. A total of 14 groundwater samples and 20 leachate samples were collected for monitoring purpose. All the samples were analyzed for various physical and chemical parameters according to standard methods: this includes pH, electrical conductivity, total dissolved solids (TDS), total hardness, and total alkalinity, major cations, major anions, and heavy metals. The results indicated that very few parameters such as pH, sulfates, and nitrates concentration in the groundwater samples are within the recommended maximum admissible limits approved by the World Health Organization and the Bureau of Indian standards. The TDS (range between 740 and 14,200 mg/L) in groundwater reveal the saline behavior of water and was found to be very high according to standards. The contour plots also indicated that the groundwater was rigorously contaminated with various heavy metals. The presence of high concentration of lead (0.59 mg/L) in groundwater samples nearby dumping site implies that groundwater samples were contaminated by leachate migration from an open dumping site.

The removal efficiencies for COD, ammoniacal nitrogen, and color, as well as ozone consumption from the Malaysian semi-aerobic landfill stabilized leachate using ozone reactor, were investigated in chapter 11. Central composite design with response surface methodology was applied to evaluate the interaction and relationship between operating variables (i.e., ozone dosage, COD concentration, and reaction time) and to develop the optimum operating condition. Based on statistical analysis, quadratic models for the four responses proven to be significant with very low probability values (<0.0001). The predicted results fit well with the results of the laboratory experiment.

The study reported in chapter 12 was conducted to investigate the treatment performance of anaerobic organism for stabilized leachate under ambient air condition. The treatability of landfill leachate was tested under various influences including pH, dosage of anaerobic organism, and contact time. Laboratory experiment revealed that anaerobic organisms were in a progressive state when the leachate was in neutral condition. At this phase, the removal efficiency of COD, ammoniacal nitrogen, and color of 76.84, 59.44, and 46.2% were experimentally attained. The experimental results also showed that the optimum variables condition was established at pH 7 of leachate sample, dosage of 100 mL of anaerobic organism. It required 14 days to reduce the COD, ammoniacal nitrogen, and color pollutants to 65.5, 60.2, and 46.3%, respectively. The preliminary

Introduction

investigation showed that the anaerobic organisms were able to remediate leachate pollutants by the microbial activity in the leachate sample under ambient air condition.

The environmental risks of sewage and leachate generation arise from it escaping into the environment, particularly to watercourses and groundwater. These risks can be mitigated by properly designed and engineered landfill sites. Further research building on these chapters is essential for tomorrow's environmental and public health.

PART I

OVERVIEW OF SEWAGE AND LEACHATE TREATMENTS

CHAPTER 1

Sustainable Treatment of Landfill Leachate

MOHAMAD ANUAR KAMARUDDIN, MOHD. SUFFIAN YUSOFF, HAMIDI ABDUL AZIZ, AND YUNG-TSE HUNG

1.1 INTRODUCTION

The exponential generation of municipal solid waste (MSW) over the years has been contributed mainly due to the expanding of industrial activities, population growth, and lifestyle changes (Ahmed and Lan 2012). In Malaysia alone, population has been increasing at a rate of 2.4 % every year and the generation of MSW also increases dramatically. As a result, various types of MSWs including industrial, commercial and agricultural byproducts are being disposed to the landfill over the years. Therefore, it is undoubtedly that appropriate MSW management is somewhat crucial (Akinbile et al. 2012) nowadays. Most significantly, Malaysians are currently generating about 5,781,600 tonnes of solid waste annually based on 2012 census data. Put together the waste generation of 0.9 kg/capita/day, it is expected that the amount of solid waste will be increased to double

---

© *2014 by the authors; licensee Springer.* Applied Water Science 5:177 (2014). DOI: 10.1007/s13201-014-0177-7. *Creative Commons Attribution license (http://creativecommons.org/licenses/by/3.0/).*

digits as the country is moving forward to be a developed nation in 2020. This estimation is by some means realistic because the process of urbanization has seen many rural and isolated areas receive widespread economic development program which has changed Malaysia landscape entirely due to the implementation of Government Transformation Program (GTP) introduced by the present 6th Malaysia's Prime Minister in 2009.

Consequently, responsible authorities particularly municipalities and landfill operators nationwide are facing difficulty in dealing with staggering amount of MSW to dispose it in a sustainable way. In addition, the selection for ideal and feasible method in controlling the disposal of high quantities of MSW at economical costs that can avoid environmental damages are difficult to be decided due to various deliberations need to be made (Umar et al. 2010). Conventionally, landfilling of solid waste has been the most preferred method for solid waste disposal due to technical feasibility, ease of operation, minimum supervisions and low operation expenditure. In most countries, landfilling is the most acceptable means for eliminating MSW which favors to the technology exploitation and capital cost (Renou et al. 2008).While most of the landfills nowadays equipped with a level three sanitary systems, many developing countries are still struggling to equip state of the art facilities at the landfill. For example, there are 261 landfills in Malaysia whereby more than 80 % of them are being controlled tipping or open dumping practice. This is due to the fact that it obscures lower cost of operation and maintenance compared to the other established techniques (incineration and advanced landfill system) (Halim et al. 2010a). Unfortunately, this practice has caused excessive generation of leachate whereby if it is not treated and safely disposed, landfill leachate could be a potential source of surface and ground water contamination, as it may percolate through soils and subsoils, causing pollution to receiving waters (Aziz et al. 2011).

The technology of solid waste disposal has evolved from conventional to advanced systems which emphasize more on the design, storage capacity and economical principle in receiving various types of wastes including leachate treatment availability. These are the main factors taken into consideration when planning a solid waste disposal site. Above all, proper decisions during designing stage, operation and long-term post-closure plan could ensure efficient monitoring of leachate generation which by far continues to generate even after the landfills have been ceased its operation (Wiszniowski et al. 2006). In general, a landfill will undergo chemical and physical changes caused from the degradation process of solid waste refuse with the soil matrix once the landfilling is complete. Generation of liquid percolates through solid waste matrix assists with rainwater percolation, biochemical, chemical and physical reactions within solid waste refuse directly influencing the quantity and quality of the leachate. In addition, leachate quality and quantity also were influenced by the landfill age, precipitation, weather variation, waste type and composition (Abbas et al. 2009). Principally, a functional landfill site is always occupied with a leachate treatment facility to treat hazardous pollutants in the leachate. Therefore, finding a sustainable method for leachate treatment has always been a priority for landfill managers in order to safely discharge treated leachate into the water bodies without endangering the environment. Over the last decades, new and advanced sustainable technologies of leachate treatment have started received growing interests which offer better removal of leachate pollutants. By utilizing these new technologies, difficult parameters are much easier to treat nowadays. In the early days, landfill leachate was mainly disposed by channeling the leachate pipes to the sewer system and released into the sea. Alternatively, there was also separated system where the leachate pipes were connected with domestic sewage network at conventional sewage plant (Ahn et al. 2002) and treated simultaneously. However, as the volume of leachate generation increase over time with wide variations in leachate pollutants, this method reduced the treatment efficiency of sewage plant (Çeçen and Aktas 2004). Concerning this, many additional treatments have been proposed and invented in treating landfill leachate separately.

Virtually, various types of treatments have been explored including biological, physical, chemical and physico-chemical techniques. As far as the authors concern, most of the treatments in the market today have their own advantages and limitations. For example, biological treatment is undoubtedly the most effective way in treating high concentration of BOD_5 (Renou et al. 2008). However, depending on the nature of leachate pollutants, sludge bulking may occur in conventional aerobic system which disturbs the leachate treatability (Dollerer and Wilderer 1996). Conventional physico-chemical techniques such as chemical precipitation

(Chen et al. 2012; Zhang et al. 2009b; Di Iaconi et al. 2010), adsorption (Ching et al. 2011; Kamaruddin et al. 2011; Lim et al. 2009; Singh et al. 2012), coagulation/flocculation (Liu et al. 2012; Al-Hamadani et al. 2011; Ghafari et al. 2010), chemical oxidation (Sun et al. 2009; Anglada et al. 2011; Cortez et al. 2011a, b) may be used as co-treatment along biological treatments. These techniques have been proven suitable in dealing with difficult parameters in leachate including humic, fulvic acid, heavy metals, adsorbable organically bound halogens (AOXs), polychlorinated biphenyls (PCBs) and several other of persistent organic pollutants (Abbas et al. 2009). Very recently, numerous studies have been introduced which focuses on new and advance treatment. In view of that, various factors have been considered in proposing an ideal treatment system that results in high efficiency of parameters reduction as to comply with the permissible discharge limit enforced by the authorities. Therefore, the purpose of this article aims to summarize leachate sustainable treatment processes including biological, physical and chemical techniques reported from 2008 to 2012. The articles discussed in depth about existing and new treatment methods in treating high concentration of leachate and its progress in the recent years.

1.2 LANDFILL LEACHATE COMPOSITION

The leachate generated from the degradation of solid wastes widely varies in terms of composition. Moreover, the risk of obtaining a concentrated leachate depends on a number of factors that control its quantity and quality, such as water percolation through the wastes, biochemical processes in wastes' cell and the degree of wastes compaction (Abbas et al. 2009; Li et al. 2010; Xu et al. 2010). Typically, leachate parameters vary depending on the age of the landfill. For instance, young leachate (1–2 years) is characterized by high organic fraction of relatively low molecular weight such as volatile organic acids, high COD, total organic carbon (TOC), BOD_5 and a $BOD_5/COD > 0.6$ (Umar et al. 2010). In contrast, old leachate (>10 years) is characterized by a relatively low chemical oxygen demand (COD) (<4,000 mg/L), slightly basic (pH > 7.5) and low biodegradability

$(BOD_5/COD <0.1)$ (Li et al. 2010). Apart from that, humic and fulvic acid and NH_3–N as well are greatly produced at this stage due to anaerobic decomposition (Bashir et al. 2011). After landfilling period, BOD_5 content will be degraded during the stabilization stage. Therefore, the BOD_5/COD ratio decreases with time because the non-biodegradable portion of COD stays unchanged in this process (Ahmed and Lan 2012). Alternatively, climate, landfill cover and type of waste at the landfill site played a major role to the leachate generation rate. A landfill site which is located at hot and arid region tends to generate smaller amount of leachate because of low precipitation whereby, leachate generation is high at tropical weather climate region due to higher precipitation infiltrates into the landfill cell (Renou et al. 2008). Utilization of cover materials during cell development whether as intermediate or final layer is one of the methods in protecting buried refuse on the landfill site to enable biodegradation of solid waste in the refuse. The utilization of impermeable type of cover materials will only increase the confining leachate amount whereby the movement of leachate within the cell is hindered and reduce the effectiveness of landfill cell. In a nut shell, having different leachate characteristics requires in depth understanding of leachate treatability to effectively reduce hazardous pollutants in leachate (Aziz et al. 2011). Table 1 shows typical leachate characteristics from semi-aerobic and anaerobic landfills in Northern Malaysia. Generally, semi-aerobic and anaerobic landfill leachate quality shows wide variation in terms of leachate parameters which indicates that aeration process plays a significant role in lowering several contaminants particularly for the case of Pulau Burung Landfill. Lower ratio of BOD_5/COD for Pulau Burung Landfill shows that the leachate is in the stabilized stage and difficult to be degraded further biologically (Aziz et al. 2010). In this case, physico-chemical process techniques are mostly recommended for stabilizing leachate (Ghafari et al. 2010). In contrast, the ratio of BOD_5/COD of 0.205 for Kulim Landfill indicates that the leachate is in the young condition and not in the stabilized stage. Previous works by various researchers (Bashir et al. 2009; Salem et al. 2008; Aghamohammadi et al. 2007) have shown that the ratio of BOD_5/COD was in the range 0.043 to 0.67 pertaining to various types of landfill leachate that are in agreement with the work by Aziz et al. (2010).

Table 1. Typical leachate characteristics from Semi-aerobic and anaerobic landfills in Malaysia.

Landfill	Semi-aerobic Pulau Burung (aerated)	Anaerobic Kulim (unaerated)	Discharge limit, DOE, Malaysia[a]
Parameters	Average values		–
Phenols (mg/L)	1.2	2.6	–
Ammonia–N	483	300	–
Total nitrogen (mg/L)	542	538	–
Nitrate–N (mg/L)	2,200	1,283	–
Nitrite–N (mg/L)	91	52	–
Total phosphorus (mg/L)	21	19	–
Orthophosphate (mg/L)	141	94	–
BOD_5 (mg/L)	83	326	50
COD	935	1,892	100
BOD5/COD	0.09	0.205	0.5
pH	8.2	7.76	5.5–9
Turbidity (NTU)	1,546	8.55	–
Color	3,334	1,936	–
Total solids (mg/L)	6,271	4,041	–
Suspended solids (mg/L)	1,437	6,336	–
Total iron (mg/L)	7.9	707	100
Zinc (mg/L)	0.6	5.3	5
Total coliform	–	0.2	1
E. Coli	–	0.81×10^{-4}	–

Adapted from Aziz et al. (2010)
[a]Second schedule (Regulation 13), amended 2013: Acceptable conditions for discharge of leachate

1.3 LEACHATE TREATMENT TECHNIQUES

Satisfactorily knowledge in landfill leachate characteristics is required to understand the variable performance found in treating the leachate either by biological, physical or physico-chemical methods. In the last few years, biological treatment has attracted more interests due to its many advantages which includes variety of sources and the ease and speed which the microorganisms can be cultured and produced (Zhao et al. 2010). These systems are divided into aerobic (with oxygen) and anaerobic (without oxygen) conditions. In particular, the use of microorganisms or bacteria to remove the contaminants in leachate is through assimilating process. This process helps to increase microbial metabolism and building blocks of the living cell. As a result, the metabolic conditions of the living cells are capable to remove leachate parameters. Regardless of the choice of application, an appropriate selection of biological treatment requires ample thought for cultivating and maintaining an acclimated healthy biomass, flow rate tolerance and organic loads to be treated. Until now, biological treatments are still one of the acceptable means in treating leachate because it offers low capital and operating cost to the operators. In addition, the application of biological treatment has been proven a total destruction of organic, sulfides, organic compounds, and toxicity.

Biological treatment has been shown very effective in removing organic and nitrogenous matter (Abbas et al. 2009) including immature leachate when the BOD_5 concentration is high and the BOD_5/COD ratio is more than 0.5 (Renou et al. 2008). However, as the biodegradation of solid waste progress, the efficiency of biological process reduces due to the increasing amount of refractory compounds namely fulvic and humic acids constituents in leachate. Nevertheless, simplicity, ease of operation and reliability have been the methods of choice in employing biological process in the early days of landfill leachate treatment process (Renou et al. 2008). In this section, we summarized a few suspended and attached growth systems that are commonly used in leachate treatment such as batch reactor, bioreactor, growth plant and microbial consortium, and combination of biological devices. These techniques, although have been seen as conventional practices, are still reliable in treating high BOD_5 contents in the landfill leachate particularly for landfill categorized as young and intermediate class. Table 2 shows some of the selection of biological treatment, their criteria and application method in a simplified format.

Table 2. Biological treatments and method of application.

Biological treatment	Common experimental condition	Example of work		
		Experimental handling	Parameters concern	Reference
Anaerobic filter/ digester/reactor	• Emit biogas (CH_4, CO_2) • Tolerable to high COD Good precipitation for toxic metals	Used seed sludge as inoculate	COD, pH, Al, Fe, Zn, Ni, Cd, Mn, Pb, Cu and Cr NH^{4+}	Kawai et al. (2012b)
		Used activated sludge as end treatment	COD, BOD_5 and TSS	Kheradmand et al. (2010)
		Co digester of leachate and sewage sludge	Biomethanation production (BMP) volatile solids reduction (VSR)	Hombach et al. (2003)
		Anaerobic sludge used as inoculums	COD, CH_4	Imen et al. (2009)
Upflow anaerobic sludge blanket (UASB)	• Normal UASB works with anaerobic bacteria	Mature leachate was co-digested with synthetic waste water	COD, CH_4	Kawai et al. (2012a)
Aerated lagoons	• Aerobic condition on top of lagoon • Anaerobic condition at the lower • High and low speed aerators used to disperse water into droplet to allow oxygen enter	Four connected aerated lagoons	COD, NH^{4+}	Mehmood et al. (2009)
Activated sludge plants/reactor	• Sludge contents is higher than aerated lagoon, possible for short residence time	Pre-denitrification activated sludge with bentonite additive	COD, NH_3–N	Wiszniowski et al. (2006)
		Phase separation through aeration	COD, BOD_5, NH^{4+} and total nitrogen	Jun et al. (2007)

Table 2. Continued.

		Utilized		
Rotating biological contactors (RBC)	• Bacteria attached to the contactors • Suitable for low organic content in leachate	Utilized single-stage anoxic RBC	NO_3^-	Cortez et al. (2011a, b)
	• Combination of reactor	Simultaneous aerobic and anaerobic (S-AA) bioreactor	COD	Yang and Zhou (2008)
	• Denitrifying reactor	Landfill simulate reactor plus activated sludge reactor	COD, NH_4^+	Shou-liang et al. (2008)
	• Reactors with denitrifying and methanogenesis	Two stage UASB and anoxic–oxic reactor	COD, BOD_5	Peng et al. (2008)
	• Partial nitrification, anaerobic ammonium oxidation (anammox) and heterotrophic denitrification	Aerobic activated sludge as inoculums and SBR as the experimental reactor	NH_4^+	Xu et al. (2010)
	• Aerobic and anaerobic condition in a reactor	Leachate recirculate plus anaerobic and aerobic of msw	pH, alkalinity, total dissolved solids, conductivity, oxidation–reduction potential, chloride, chemical oxygen demand, ammonia, and total Kjeldahl nitrogen, in addition to generated leachate quantity	Bilgili et al. (2007)
Biological co-treatment	• Selection of disc for cyclic bath RBC	RBC and upward-flow anaerobic sludge bed reactor	COD	Castillo et al. (2007)
	• Different hydraulic retention times (HRT), rotational speeds, and with varying organic concentrations			

1.3.1 BIOLOGICAL PROCESS

1.3.1.1 BATCH REACTOR

Xu et al. (2010) performed a partial nitrification, aerobic ammonium oxidation (Anammox) and heterothopic denitrification by sequencing batch reactor (SBR). The experimental conditions of 30 ± 1 °C and dissolved oxygen (DO) of concentration within 1.0–1.5 mg/L were fixed in the SBR. They found that maximum aerobic ammonium oxidizing and anaerobic ammonium oxidizing are achieved at 0.79 and 0.18 (kg –N/kg$_{dw}$/day) after the inoculation of Anammox biomass and aerobic activated sludge (80 % w/w) that last for 86 days In contrast, aerobic ammonium oxidizing, anaerobic ammonium oxidizing and denitrification reached 2.83, 0.65 and 0.11 (kg –N/kgdw/day) when denitrifying bacteria was inoculated into the reactor along with the feeding of raw landfill leachate. In other study, Spagni and Marsili-Libelli (2009) focused on the nitritation and denitritation processes of stabilized leachate by SBR process to enhance the nitrogen removal efficiency. They reported that by adding external COD and adjusting the length of oxic phase could increase nitrogen rate removal. Meanwhile, Lan et al. (2011) successfully conducted simultaneous partial nitrification anammox and denitrification (SNAD) process by SBR which focused on the influence of hydraulic retention time (HRT). They concluded that increasing the HRT from day 3 to 9 of SBR process would increase the COD (87–96 %). Meanwhile, different observations were recorded when pH and DO were reduced which result in lower removal of COD and nitrogen. Finally, they revealed total nitrogen (TN) removal of 85–87 % by anammox with partial nitrification and 7–9 % by denitrification from the SNAD process, respectively. Aziz et al. (2011) utilized SBR instruments for the swim-bed biofringe process for the removal of COD, BOD$_5$, TKN and NH$_3$–N from stabilized leachate. They utilized activated sludge and biofringe as the main process parameters. The results demonstrated that swim-bed BF was capable of removing nitrite, nitrate and phosphorus from leachate. On the contrary, the removal performance for COD and NH$_3$–N was not significant, respectively.

Sustainable Treatment of Landfill Leachate 13

1.3.1.2 BIOREACTOR

Yahmed et al. (2009) conducted an investigation of a pilot unit system consisting of three unit fixed bioreactors. They tested for different organic loading rate (OLR) of microbials namely *Actinomycetes*, *Bacillus*, *Pseudomonas* and *Burkholderia* for the removal of TOC. They concluded that the maximum TOC reduction by *Pseudomonas* isolates was of 70 %. Meanwhile, *Actinomycetes* isolates, *Bacillus* isolates and *Burkholderia* isolates gave 69, 69 and 77 % TOC reduction, respectively. In another study, Ellouze et al. (2008) investigated leachate treatability by utilizing sludge from a waste water treatment plant. Preliminary studies showed that the acclimatization of the sludge was able to remove organic matter and toxicity. A set up of stirred tank reactor with OLR from 0.5 to 4 g/L/day with HRT decreased from 50 to 4.6 days demonstrating that COD was removed up to 80 % for a loading rate of 5.4 g/L/day. In addition, the concentration of $N-NH^{4+}$ was reduced below to the recommended standard. Finally, the results from toxicity of *Vibrio fischeri* and the germination of *Lepidium sativum* seeds showed that the treatment was able to effectively provide detoxification of the effluent whereby the loading rate up to 6 g/L was ideal for the perturbation of the system which triggered an accumulation of residual COD and toxicity, respectively. Ismail et al. (2011) investigated the effect of different organic loading charges (0.6–16.3 kg) for the removal of TOC and TKN by submerged biofilm reactor. The results showed that without initial pH adjustment, TOC removal rate varied between 65 and 97 %. The total reduction of COD reached 92 % at a HRT of 36 h. However, the removal of total Kjeldahl nitrogen for loading charges of 0.5 kg $N/m^3/day$ reached 75 %. Further toxicity test for the removal of organic carbon and nitrogen showed that *Bacillus*, *Actinomyces*, *Pseudomonas* and *Burkholderia genera* were responsible for these occurrences. Chen et al. (2008) investigated the performance of a moving bed biofilm reactor (MMBR) via aerobic and anaerobic sequence for simultaneous removal of COD and ammonium. They discovered that anaerobic MBBR played a major role in COD removal (91 %) at OLR of 4.08 kg $COD/m^3/day$ due to methanogenesis and the aerobic MBBR acted as COD-polishing and ammonium removal step. In contrast, HRT at 1.25 days required to remove more than 97 % of NH^{4+} of the aerobic MMBR. Bohdziewicz et al. (2008)

examined the treatability of leachate by submerged membrane bioreactors. They used synthetic waste water as feeding medium by volume ratio with the addition of leachate dilution between 50 and 75 %. They claimed that higher COD removal could be achieved with the leachate addition of 10–20 % v/v. They also revealed that the best anaerobic digestion efficiency (COD removal 90 %) was observed for HRT for 2 days and OLR of 2.5 kg COD/m³ days for the optimal anaerobic digestion efficiency.

1.3.2 GROWTH PLANT AND MICROBIAL CONSORTIUM

Ye et al. (2008) tested immobilized microbial for the removal of COD and NH_3–N. They measured the efficient microbial flora on the carrier by Kjeldahl's method. The biological process showed that immobilized microorganisms system was effective for the removal of COD and nitrogen at 98.3 and 99.9 %, respectively. A study done by Saetang and Babel (2012) revealed that *Trametes versicolor* BCC 8725 could remove 78 color, 68 BOD_5 and 57 % COD from leachate sample within 15 days at optimum condition, respectively. They also claimed that organic loading and ammonia were the factors that affected the biodegradation. In another work, Białowiec et al. (2012) compared reed and willow with an unplanted control by measuring redox potential levels in the rhizosphere of microcosm system for the leachate bioremediation. The results suggested that redox potential in the reed *rhizosphere* was anoxic (mean −102 ± 85 mV), but it was the least negative, being significantly higher than in the willow (mean −286 ± 118 mV), which had the lowest Eh. They also reported that NH^{4+} reduced from the first day and remained at a similar low level until 4 weeks of the experimental period. Meanwhile, Loncnar et al. (2010) discovered that the planted willows at a recirculation process of leachate showed a high sustainability of saline ions. The concentration of saline ions was recorded at ranges 132 to 2,592 mg Cl^-/L, 69 to 1,310 mg Na^+/L and 66 to 2,156 mg K^+/L, with mean values of 1,010, 632 and 686 mg/L, respectively. Akinbile et al. (2012) found that by utilization of *Cyperus haspan* with sand and gravel in a constructed wetland with optimum retention time of 3 weeks could efficiently reduce heavy metals parameters at the ranges of 33–89 %. Meanwhile, significant reduction of TSS, COD,

Sustainable Treatment of Landfill Leachate

BOD$_5$, NH$_3$–N, and TP of 98, 92, 79, 54 and 99 % was recorded, respectively. In another work, using anaerobic organisms in a series of anaerobic tanks filled with leachate, 100 mL of anaerobic organism and 14 days of microbial inhibitors, 65.5, 60.2 and 46.3 % of COD, NH$_3$–N and color were removed, respectively (Kamaruddin et al. 2013).

1.3.3 PHYSICAL–CHEMICAL PROCESS

Generally, satisfactory treatment of landfill leachate is dependent on methods applied to leachate generation handling. A complete landfill leachate treatment usually consists of physical, chemical and biological processes. Physical treatment utilizes non chemical or biological changes in the leachate whereby only physical phenomenon is used to enhance leachate quality. For example, screening of leachate is done by employing metal grit trap to retain larger impurities prior to subsequent treatment. Meanwhile, sedimentation process is involving settling of solids by gravitational force by simply allowing short residence time in sedimentation tank. This process is crucial for flocs formation. Another type of physical treatment is aeration which utilizes oxygen as the oxidation agent in leachate lagoon. This process has been found to enhance the removal of BOD$_5$ in pre-treatment as proven by many successful treatment selections. In contrast, chemical treatment utilizes chemicals additive that involves reaction to improve leachate quality. During chemical treatment, neutralization is commonly used to neutral leachate condition by the addition of acid or base in the process. In other process, coagulation has been known as one of the oldest chemical treatment in landfill leachate treatment. It utilizes chemical additives which enable the formation of insoluble end products and capable of removing a wide range of leachate parameters through ionic mechanism. In addition, certain types of polyvalent metals are widely used as coagulant or coagulant aid such as ferric chloride, polyaluminum chloride, aluminum sulfate or ferric sulfate. Alternatively, disinfection of leachate is one of established methods in chemical treatment. Chlorine known as the strong oxidizing agent is commonly used to kill bacteria when crucial biological process is affected by the chlorine. In a nut shell, physical–chemical process, includes adsorption, coagulation/

16   Sewage and Landfill Leachate

flocculation and chemical oxidation, is commonly used when the biological process is hindered due to excessive presence of refractory compounds in leachate. Normally, physical–chemical process is carried out as a pretreatment or at the final stage of the leachate treatment process. Table 3 discusses the criteria of the most common biological and physical–chemical process in leachate treatment and their advantages.

1.3.3.1 ACTIVATED CARBON ADSORPTION

Adsorption of leachate by activated carbon has received great interests considering its superior properties having larger surface area, high adsorption capacity and better thermal stability. Ching et al. (2011) used a chemically treated coffee ground-activated carbon for the removal of total iron and orthophosphate from stabilized leachate. They discovered that optimum removal for the latter was attained at impregnation ratios (IRs) of 2.5 and 0.5 at doses of 10 g and pH 8.1. In contrast, pH 13 was found optimum for total iron removal while pH < 5 and >11 was optimum for PO_4–P removal. Kamaruddin et al. (2011) concluded that the optimum preparation conditions of durian peel-activated carbon (DPAC) was achieved at IR, activation temperature, and activation time of 3, 400 °C and 2.2 h, for the removal of NH_3–N from stabilized leachate. The optimum conditions of DPAC are capable of removing 47 % of NH_3–N. Kalderis et al. (2008) investigated $ZnCl_2$-treated rice husk and sugarcane bagasse-activated carbons. The activated carbons were tested for humic acid, phenol and leachate parameters removal. They found that both ACs showed the best adsorption behavior towards phenol, removing around 80 % at 4 h equilibrium period. However, the adsorption for arsenic and humic acids was lower than that of phenol based on isotherm data. Finally, they revealed that with 30 g/L of AC, it was possible to remove 70 and 60 % of COD and color, respectively. Singh et al. (2012) developed isotherm and kinetic models for three types of commercially available activated carbons. They suggested that Redlich–Peterson model showed better fit to the experimental data and the TOC adsorption capacity for both micro-porous and meso-porous activated carbons. In addition, intraparticle diffusion coefficients (De) for both AC were in the order 10^{-10} m²/s for particle sizes

Sustainable Treatment of Landfill Leachate

Table 3. Criteria of biological and physical–chemical treatment.

Treatment option	Treatment process	Treatment efficiency			Operational cost	Space requirement
		Leachate condition				
		Young leachate	Medium age leachate	Mature leachate		
Biological	Rotating biological contactor (RBC)	Strong	Fair	Weak	Expensive	Normal
	Sequencing batch reactor (SBR)	Strong	Fair	Weak	Moderate	Normal
	Moving bed biofilm reactor (MBBR)	Strong	Fair	Weak	Expensive	Large
	Membrane bioreactor (MBR)	Strong	Fair	Weak	Expensive	Large
	Upflow anaerobic sludge blanket (UASB)	Strong	Fair	Fair	Moderate	Normal
	Activated sludge	Strong	Fair	Weak	Expensive	Large
	Phytoremediation	Fair	Fair	Good	Inexpensive	Large
	Lagooning	Strong	Fair	Weak	Expensive	Large
Physical-chemical	Adsorption	Weak	Fair	Weak	Expensive	Normal
	Coagulation	Weak	Fair	Fair	Inexpensive	Medium
	Chemical oxidation	Weak	Fair	Fair	Expensive	Normal
	Stripping	Weak	Fair	Fair	Expensive	Large
	Precipitation	Weak	Fair	Fair	Inexpensive	Medium

>0.5 mm. Lim et al. (2009) established an axial dispersion model for palm shell-activated carbon (PSAC) in column mode. The applicability of the model was tested for the removal of COD and turbidity of leachate. The highest breakthrough of COD was obtained at Empty Bed Contact Time (EBCT) of 14.7 min, with sorption capacity of 1,460 mg/g. In contrast, turbidity and pH effluent showed insignificant effect on EBCT, respectively.

While activated carbon has gained much popularity in the market nowadays, there is also several type of adsorbents receiving great interest in the recent years due to their abundance, easily obtained, high regeneration cycle, and higher mechanical stability in adsorption studies. Accordingly, waste materials such as from agricultural sectors and industrial byproducts have been identified to have the potential as an alternative adsorbent in adsorption studies. Table 4 shows several types of adsorbents that have been proposed and tested in treating landfill leachate by adsorption studies.

1.3.3.2 COAGULATION FLOCCULATION

Coagulation and flocculation is known as one of the oldest treatment methods in landfill leachate. Apart from that, it has been widely used in treating stabilized (Al-Hamadani et al. 2011) and matured landfill leachate (Vedrenne et al. 2012). In addition, the application of coagulation and flocculation can be used as pre-treatment process in order to remove non-biodegradable organic matter (Renou et al. 2008). Several studies have identified the selection of appropriate experimental conditions when employing coagulant and flocculation process. Ghafari et al. (2009) used PAC and alum to treat stabilized leachate in coagulation/flocculation process at maintained mixing time and mixing speed. They utilized CCD and RSM to establish the relationship between operating variables (dosage and pH) and leachate parameters removal. The results indicated that the optimum conditions for PAC was obtained at dosage of 2 g/L and ph 7.5 which managed to reduce COD, turbidity, color and TSS concentrations at 43.1, 94.0, 90.7, and 92.2 %. Subsequently, the optimum condition for alum was achieved at dosage 9.5 g/L and pH 7 which further reduced COD, turbidity, color and TSS concentrations to 62.8, 88.4, 86.4, and 90.1 % respectively. However, when they optimized the speed and time for rapid

Sustainable Treatment of Landfill Leachate

Table 4. Typical leachate characteristics from Semi-aerobic and anaerobic landfills in Malaysia.

Adsorbent	Source	Parameters concern	Reference
Turkish clinoptilolite	Local supply	Ammonium	Karadag et al. (2008)
Ion resins	Local supply	Color, COD, NH_3–N,	Bashir et al. (2010)
Kemiron	Local supply	Arsenic	Oti et al. (2011)
Honeycomb cinders	Byproducts from briquette combustion	PO_4–P, COD	Yue et al. (2011)
Sphagnum peat moss	Local supply	Cd, Ni	Champagne and Li (2009)
Crushed mollusk shells	Local supply	Cd, Ni	
Composite adsorbent	Local supply and Agri-wastes	NH_3–N, COD	Halim et al. (2010b)
Limestone, granular AC	Local supply	Orthophosphate	Hussain et al. (2011)
Activated carbon, bone meal and iron fines	Local supply	Al, As, Ca, Cd, Co, Cr, Cu, Fe, Hg, Mg, Mn, Mo, Ni, Pb, Sr and Zn	Modin et al. (2011)
Coal fly ash	Thermal power plant	Zn, Pb, Cd, Mn and Cu	Mohan and Gandhimathi (2009)
Durio zibethinus L.	Agricultural waste	NH_3–N, carbon yield	Kamaruddin et al. (2011)

and slow mixing, they observed that COD removal was achieved at 84.5 and 56.7 % for alum and PAC. Single use of PAC showed that turbidity, 99.18 %; color, 97.26 % and TSS, 99.22 % were achieved; whereas alum showed inferior removal (turbidity, 94.82 %; color, 92.23 % and TSS, 95.92 %) (Ghafari et al. 2010). Liu et al. (2012) used RSM for the optimization process of polyferricsulphate (PFS) coagulant towards COD, color, turbidity and HA removal. At optimum conditions, COD, color, turbidity and HA removal of 56.38, 63.38, 89.79, 70.41 % were observed at PFS dose of 8 g/L at pH 6.0, $FeCl_3 \cdot 6H_2O$ dose of 10 g/L at pH 8.0 and $Fe_2(SO_4)_3 \cdot 7H_2O$ dose of 12 g/L at pH 7.5. Using similar optimum variable conditions, 68.65, 93.31, 98.85, 80.18 % for $FeCl_3 \cdot 6H_2O$ and 55.87, 74.65, 94.13, 53.64 % for $Fe_2(SO_4)_3 \cdot 7H_2O$ of COD_{cr}, color, turbidity and

HA removal were observed, respectively. In another study, an alternative coagulant was successfully synthesized and tested. Al-Hamadani et al. (2011) compared psyllium husk as coagulant aid with PACl and alum. They found that the maximum removal was achieved when psyllium husk was used as coagulant aid with PACl resulting in COD, color and TSS removal of 64, 90 and 96 %, respectively. Meanwhile, Syafalni et al. (2012) compared lateritic soil coagulant with alum in jar test experiment. The optimum condition was achieved at pH 2 and lateritic soil coagulant dose of 14 g/L resulting 65.7 % COD, 81.8 % color and 41.2 % NH_3–N removal. Comparable finding was observed when alum was used at pH 4.8 and coagulant dosage of 10 g/L where COD, color and NH_3–N were removed at 85.4, 96.4 and 47.6 %, respectively. Tzoupanos et al. (2008) evaluated the performance of polyaluminium silicate chloride (PSC) coagulant with different Al to Si molar ratio with biologically treated leachate. The results suggested that PSC had better removal of COD and color than PACl due to high tolerance against pH ranging from 7 to 9. Concerning with the inhibitory of dissolved organic matters, Comstock et al. (2010) compared three types of coagulants which focused on dissolved organic matter (DOM) removal from leachate. The presence of DOM was measured using specific ultraviolet (UV) absorbance at 254 nm ($SUVA_{254}$) and fluorescence excitation–emission matrices. The performance of the metals salts was in the order of: ferric sulfate > aluminum sulfate > ferric chloride and DOM removal followed the trend of color > UV_{254} > dissolved organic carbon > COD. In another study, Yimin et al. (2008) used poly-magnesium–aluminum sulfate (PMAS) in jar test experiment. The removal of COD, BOD_5, UV_{254},(OM) by PMAS was observed at 65, 60, 85 % under optimum conditions, respectively.

1.3.3.3 CHEMICAL OXIDATION

Generally, chemical oxidation process utilizes chemical substances, mainly chlorine, ozone, potassium permanganate and calcium hydroxide (Abbas et al. 2009). In addition, advance oxidation process (AOP) normally is used to enhance the chemical oxidation efficiency to the stable oxidation state. Owing to the successful rate of the removal of refractory

Sustainable Treatment of Landfill Leachate

compounds in leachate, AOP, however, has some limitation including high energy requirement, and chemical reagent (Kalderis et al. 2008) throughout the leachate treatment process. Nevertheless, AOP still considered as the better treatment methods when employing it as pre-treatment prior to the biological process thereby reducing capital operation of leachate treatment. Previous studies have demonstrated that chemical precipitation, Fenton/Electro-Fenton/Photo-Fenton, Photochemical/Photoelectrochemical/Photocatalytic could significantly reduce leachate containing refractory compounds. These processes include both non photochemical and photochemicals which generate hydroxyl radicals with and without light energy (Wiszniowski et al. 2006). Table 5 summarizes some of the major breakthroughs in the utilization of AOP techniques which results in significant removal of leachate pollutants.

1.3.4 ADVANCED BIOLOGICAL/PHYSICAL–CHEMICAL TECHNIQUES

With stringent requirement by authorities in protecting environmental fate, the treatability of landfill leachate is a prominent challenge for the landfill operator to comply with the current regulations. With regards to this, conventional treatment is not sufficient to render high concentration of leachate pollutants. Therefore, the adverse impacts of inefficient leachate treatment have raised serious concerns to the society and environment, respectively. Ultimately, the combination of individual treatment process into hybrid process has been more effective and emerged as the choice of treatment for landfill operators. Kwon et al. (2008) found that higher reduction of COD_{cr}, color and TP could be achieved when they employed nanofiltration-rotary disk membrane (NF-RDM) process. In addition, the introduction of RO with NF-RDM process enhanced NH^{4+} removal from 25 to 92 %. In another study, Tsilogeorgis et al. (2008) concluded that ultrafiltration membrane-SBR was able to remove TN removal (88 % maximum) over 4 months monitoring. However, COD removal varied (40–60 %) due to high SRT. Also, PO_4–P removal efficiency was varied (35–45 %) during the first 50 days of operation due to direct addition of KH_2PO_4/K_2HPO_4 that was aimed to improve C:N:P ratio.

In a hybrid experimental work, Li et al. (2010) investigated coagulation/flocculation augmented powdered activated carbon (PAC). They used four types of commercially available coagulants to determine optimum working conditions and found that PFS showed better removal for COD, SS, turbidity, toxicity and sludge volume at 70, 93, 97 % and 32 mL. Consequently, 10 g/L of PAC was found optimum with 90 min contact time during experimental period. Under optimum conditions of combined techniques, COD, Pb, Fe and toxicity removals were found 86, 97.6, 99.7 and 78 %, respectively. Meanwhile, to improve pollutants removal, Palaniandy et al. (2010) found that the combination between $FeCl_3$ coagulation and dissolved air flotation (DAF) managed to reduce turbidity, COD, color and NH_3–N concentration up to 50, 75, 93 and 41 %. The statistical analysis suggested that the optimum operating conditions for coagulation and DAF were 599.22 mg/L of $FeCl_3$ at pH 4.76 followed by saturator pressure of 600 kPa, flow rate of 6 L/min and injection time of 101 min. In another work, Poznyak et al. (2008) injected ozone process after the coagulation/flocculation treatment. They found that coagulation/flocculation injected ozone could remove 70 % of humic substances in leachate. Next, when ozone process was further induced, color was 100 % removed during 5 min period. Finally, they found that organic substance diminished completely during 15 min ozonation when extracted with chloroform–methanol and 5 min when extracted with benzene. Ying et al. (2012a) applied various treatment processes with combination of internal micro-electrolysis (IME) without aeration and IME with full aeration in one reactor. The authors implemented a novel sequencing batch internal micro-electrolysis reactor (SIME) throughout the experimental work. Results showed that high COD removal efficiency of 73.7 ± 1.3 % was obtained which was 15.2 and 24.8 % higher than that of the IME with and without aeration, respectively. The SIME reactor also exhibited a COD removal efficiency of 86.1 ± 3.8 % to mature landfill leachate in the continuous operation, which was much higher ($p < 0.05$) than that of conventional treatments of electrolysis (22.8–47.0 %), coagulation–sedimentation (18.5–22.2 %), and the Fenton process (19.9–40.2 %), respectively (Ying et al. 2012b).

Among advanced oxidation processes, several improvements towards the capabilities of existing techniques have been explored by various authors. Galeano et al. (2011) utilized catalytic wet peroxide oxidation (CWPO) with an Al/Fe-pillared clay catalyst in semi-batch reactor. The COD was found reduced up to 50 % and biodegradability index (BI) output was exceeding 0.3 during 4-h experiment duration. They concluded that high catalyst, low peroxide concentrations, dosages and addition rates were the main factors affecting oxidizing agents in terms of BI and COD removal efficiency. Xu et al. (2012) found that by applying catalytic wet air oxidation (CWAO) with the presence of AC as catalyst and potassium persulfate ($K_2S_2O_8$) as promoter, almost complete fulvic acid (FA) and COD removal up to 78 % could be achieved in the $K_2S_2O_8$/AC system at 150 °C and 0.5 MPa oxygen pressure. They also found that the BOD_5/COD ratio increased from 0.13 to 0.95 after CWAO. Sun et al. (2009) compared the application of Fenton and Oxone/Co^{2+} oxidation processes. When they tested Fenton oxidation as standalone process, COD removal was found at 56.9 % but SS and color increased in concentration due to high generation of ferric hydroxide sludge. Subsequently, when they assessed the performance of Oxone/Co^{2+} oxidation, the removal of COD, SS and color removal increased to 57.5, 53.3 and 83.3 %. The optimum conditions of the process were: [Oxone] = 4.5 mmol/L, [Oxone]/[Co^{2+}] = 104, pH = 6.5, reaction temperature = 30 ± 1 °C, reaction time = 300 min, number of stepwise addition = 7. Panizza et al. (2010) utilized anodic oxidation using electrolyte flow cell equipped with lead dioxide (PbO_2) anode and stainless steel as cathode. They observed that the galvanostatic electrolyses enhanced COD removal along with rising current, solution pH and temperature. Gabarró et al. (2012) studied the effects of temperature on NH_3–N in a partial nitration (PN)-SBR. The stable PN was achieved with minimum volume of 111 L and N–NH^{4+} of 6,000 mg/L at 25 and 35°C. The result was demonstrated by kinetic model where NH^{4+} and NO_2 concentrations were similar at both temperatures. In contrast, free ammonia and free nitrous acid (FNA) were found differed due to the strong temperature dependence. There are concerns with excessive pollutants concentration in matured leachate,

1.4 CONCLUSIONS

Over the years, various sustainable landfill leachate treatment techniques have been proposed and tested for treating highly polluted leachate. At this point, here are some of the key points from the extensive discussions regarding sustainable landfill leachate treatment:

- Refractory compounds in leachate always change over times due to overwhelmed mankind activities. Therefore, modification of existing treatment technique may be viable to ensure that the treatment efficiency is consistent and in accordance to the regulatory standards;
- there has been a steady progress of new and advanced sustainable landfill leachate treatment which proven to be a promising alternative;
- utilization of advanced waste disposal method such as incineration and recycling may be suitable to mitigate the generation of landfill leachate.
- Though there are still uncertainties whether these techniques could enhance environmental sustainability and safety of human being, more efforts should be carried out to ensure a livelihood of human being and earth coexistence;
- therefore, a holistic approach is essential for finding a suitable leachate treatment opportunity in order to safeguard environmental and human being livelihood, as a whole.

REFERENCES

1. Abbas AA, Jingsong G, Ping LZ, Pan YY, Al-Rekabi WS (2009) Review on landfill leachate treatments. J Appl Sci Res 5:534–545
2. Aghamohammadi N, Aziz HBA, Isa MH, Zinatizadeh AA (2007) Powdered activated carbon augmented activated sludge process for treatment of semi-aerobic landfill leachate using response surface methodology. Bioresour Technol 98:3570–3578
3. Ahmed FN, Lan CQ (2012) Treatment of landfill leachate using membrane bioreactors: a review. Desalination 287:41–54

Sustainable Treatment of Landfill Leachate

4. Ahn WY, Kang MS, Yim SK, Choi KH (2002) Advanced landfill leachate treatment using an integrated membrane process. Desalination 149:109–114

5. Akinbile CO, Yusoff MS, Ahmad Zuki AZ (2012) Landfill leachate treatment using sub-surface flow constructed wetland by Cyperus haspan. Waste Manage 32:1387–1393

6. Al-Hamadani YAJ, Yusoff MS, Umar M, Bashir MJK, Adlan MN (2011) Application of psyllium husk as coagulant and coagulant aid in semi-aerobic landfill leachate treatment. J Hazard Mater 190:582–587

7. Anglada Á, Urtiaga A, Ortiz I, Mantzavinos D, Diamadopoulos E (2011) Boron-doped diamond anodic treatment of landfill leachate: evaluation of operating variables and formation of oxidation by-products. Water Res 45:828–838

8. Atmaca E (2009) Treatment of landfill leachate by using electro-Fenton method. J Hazard Mater 163:109–114

9. Aziz SQ, Aziz HA, Yusoff MS, Bashir MJK, Umar M (2010) Leachate characterization in semi-aerobic and anaerobic sanitary landfills: a comparative study. J Environ Manag 91:2608–2614

10. Aziz HA, Ling TJ, Haque AAM, Umar M, Adlan MN (2011) Leachate treatment by swim-bed bio fringe technology. Desalination 276:278–286

11. Bashir MJK, Isa MH, Kutty SRM, Awang ZB, Aziz HA, Mohajeri S, Farooqi IH (2009) Landfill leachate treatment by electrochemical oxidation. Waste Manag 29:2534–2541

12. Bashir MJK, Aziz HA, Yusoff MS, Aziz SQ, Mohajeri S (2010) Stabilized sanitary landfill leachate treatment using anionic resin: treatment optimization by response surface methodology. J Hazard Mater 182:115–122

13. Bashir MJ, Aziz HA, Yusoff MS (2011) New sequential treatment for mature landfill leachate by cationic/anionic and anionic/cationic processes: optimization and comparative study. J Hazard Mater 186:92–102

14. Białowiec A, Davies L, Albuquerque A, Randerson PF (2012) Nitrogen removal from landfill leachate in constructed wetlands with reed and willow: redox potential in the root zone. J Environ Manag 97:22–27

15. Bilgili MS, Demir A, Özkaya B (2007) Influence of leachate recirculation on aerobic and anaerobic decomposition of solid wastes. J Hazard Mater 143:177–183

16. Bohdziewicz J, Neczaj E, Kwarciak A (2008) Landfill leachate treatment by means of anaerobic membrane bioreactor. Desalination 221:559–565

17. Castillo E, Vergara M, Moreno Y (2007) Landfill leachate treatment using a rotating biological contactor and an upward-flow anaerobic sludge bed reactor. Waste Manag 27:720–726

18. Çeçen F, Aktas Ö (2004) Aerobic co-treatment of landfill leachate with domestic wastewater. Environ Eng Sci 21:303–312

19. Champagne P, Li C (2009) Use of Sphagnum peat moss and crushed mollusk shells in fixed-bed columns for the treatment of synthetic landfill leachate. J Mater Cycles Waste 11:339–347

20. Chen S, Sun D, Chung JS (2008) Simultaneous removal of COD and ammonium from landfill leachate using an anaerobic–aerobic moving-bed biofilm reactor system. Waste Manag 28:339–346

21. Chen YN, Liu CH, Nie JX, Luo XP, Wang DS (2012) Chemical precipitation and biosorption treating landfill leachate to remove ammonium-nitrogen. Clean Technol Envir 1–5

22. Ching SL, Yusoff MS, Aziz HA, Umar M (2011) Influence of impregnation ratio on coffee ground activated carbon as landfill leachate adsorbent for removal of total iron and orthophosphate. Desalination 279:225–234

23. Comstock Seh, Boyer TH, Graf KC, Townsend TG (2010) Effect of landfill characteristics on leachate organic matter properties and coagulation treatability. Chemosphere 81:976–983

24. Cortez S, Teixeira P, Oliveira R, Mota M (2011a) Evaluation of Fenton and ozone-based advanced oxidation processes as mature landfill leachate pre-treatments. J Environ Manag 92:749–755

25. Cortez S, Teixeira P, Oliveira R, Mota M (2011b) Mature landfill leachate treatment by denitrification and ozonation. Process Biochem 46:148–153

26. Di Iaconi C, Pagano M, Ramadori R, Lopez A (2010) Nitrogen recovery from a stabilized municipal landfill leachate. Bioresour Technol 101:1732–1736

27. Dollerer J, Wilderer P (1996) Biological treatment of leachates from hazardous waste landfills using SBBR technology. Water Sci Technol 34:437–444

28. Ellouze M, Aloui F, Sayadi S (2008) Performance of biological treatment of high-level ammonia landfill leachate. Environ Technol Environ Technol 29:1169–1178

29. Gabarró J, Ganigué R, Gich F, Ruscalleda M, Balaguer M, Colprim J (2012) Effect of temperature on AOB activity of a partial nitritation SBR treating landfill leachate with extremely high nitrogen concentration. Bioresour Technol 126:283–289

30. Galeano LA, Vicente MÁ, Gil A (2011) Treatment of municipal leachate of landfill by Fenton-like heterogeneous catalytic wet peroxide oxidation using an Al/Fe-pillared montmorillonite as active catalyst. Chem Eng J 178:146–153

31. Ghafari S, Aziz HA, Isa MH, Zinatizadeh AA (2009) Application of response surface methodology (RSM) to optimize coagulation–flocculation treatment of leachate using poly-aluminum chloride (PAC) and alum. J Hazard Mater 163:650–656

32. Ghafari S, Aziz HA, Bashir MJK (2010) The use of poly-aluminum chloride and alum for the treatment of partially stabilized leachate: a comparative study. Desalination 257:110–116

33. Gunay A, Karadag D, Tosun I, Ozturk M (2008) Use of magnesit as a magnesium source for ammonium removal from leachate. J Hazard Mater 156:619–623

34. Halim AA, Aziz HA, Johari MAM, Ariffin KS (2010a) Comparison study of ammonia and COD adsorption on zeolite, activated carbon and composite materials in landfill leachate treatment. Desalination 262:31–35

35. Halim AA, Aziz HA, Johari MAM, Ariffin KS, Adlan MN (2010b) Ammoniacal nitrogen and COD removal from semi-aerobic landfill leachate using a composite adsorbent: fixed bed column adsorption performance. J Hazard Mater 175:960–964

36. Hermosilla D, Cortijo M, Huang CP (2009) Optimizing the treatment of landfill leachate by conventional Fenton and photo-Fenton processes. Sci Total Environ 407:3473–3481

37. Hombach ST, Oleszkiewicz JA, Lagasse P, Amy LB, Zaleski AA, Smyrski K (2003) Impact of landfill leachate on anaerobic digestion of sewage sludge. Environ Technol 24:553–560

Sustainable Treatment of Landfill Leachate

38. Hussain S, Aziz HA, Isa MH, Ahmad A, Van Leeuwen J, Zou L, Beecham S, Umar M (2011) Orthophosphate removal from domestic wastewater using limestone and granular activated carbon. Desalination 271:265–272
39. Imen S, Ismail T, Sami S, Fathi A, Khaled M, Ahmed G, Latifa B (2009) Characterization and anaerobic batch reactor treatment of Jebel Chakir Landfill leachate. Desalination 246:417–424
40. Ismail T, Tarek D, Mejdi S, Amira BY, Murano F, Neyla S, Naceur J (2011) Cascade bioreactor with submerged biofilm for aerobic treatment of Tunisian landfill leachate. Bioresour Technol 102:7700–7706
41. Jia C, Wang Y, Zhang C, Qin Q (2011) UV-TiO2 photocatalytic degradation of landfill leachate. Water Air Soil Pollut 217:375–385
42. Jun D, Yongsheng Z, Henry RK, Mei H (2007) Impacts of aeration and active sludge addition on leachate recirculation bioreactor. J Hazard Mater 147:240–248
43. Kalderis D, Koutoulakis D, Paraskeva P, Diamadopoulos E, Otal E, Valle JOD, Fernández-Pereira C (2008) Adsorption of polluting substances on activated carbons prepared from rice husk and sugarcane bagasse. Chem Eng J 144:42–50
44. Kamaruddin MA, Yusoff MS, Ahmad MA (2011) Optimization of durian peel based activated carbon preparation conditions for ammoniacal nitrogen removal from semi-aerobic landfill leachate. J Sci Ind Res 70:554–560
45. Kamaruddin MA, Yusoff MS, Aziz HA, Basri NK (2013) Removal of COD, ammoniacal nitrogen and colour from stabilized landfill leachate by anaerobic organism. Appl Water Sci 3:359–366
46. Karadag D, Tok S, Akgul E, Turan M, Ozturk M, Demir A (2008) Ammonium removal from sanitary landfill leachate using natural Gördes clinoptilolite. J Hazard Mater 153:60–66
47. Kawai M, Kishi M, Hamersley MR, Nagao N, Hermana J, Toda T (2012a) Biodegradability and methane productivity during anaerobic co-digestion of refractory leachate. Int Biodeterior Biodegrad 72:46–51
48. Kawai M, Purwanti IF, Nagao N, Slamet A, Hermana J, Toda T (2012b) Seasonal variation in chemical properties and degradability by anaerobic digestion of landfill leachate at Benowo in Surabaya, Indonesia. J Environ Manag 110:267–275
49. Kheradmand S, Karimi-Jashni A, Sartaj M (2010) Treatment of municipal landfill leachate using a combined anaerobic digester and activated sludge system. Waste Manag 30:1025–1031
50. Kwon O, Lee Y, Noh S (2008) Performance of the NF-RDM (rotary disk membrane) module for the treatment of landfill leachate. Desalination 234:378–385
51. Lan CJ, Kumar M, Wang CC, Lin JG (2011) Development of simultaneous partial nitrification, anammox and denitrification (SNAD) process in a sequential batch reactor. Bioresour Technol 102:5514–5519
52. Li W, Hua T, Zhou Q, Zhang S, Li F (2010) Treatment of stabilized landfill leachate by the combined process of coagulation/flocculation and powder activated carbon adsorption. Desalination 264:56–62
53. Lim YN, Shaaban MG, Yin CY (2009) Treatment of landfill leachate using palm shell-activated carbon column: axial dispersion modeling and treatment profile. Chem Eng J 146:86–89

54. Liu X, Li XM, Yang Q, Yue X, Shen TT, Zheng W, Luo K, Sun YH, Zeng GM (2012) Landfill leachate pretreatment by coagulation–flocculation process using iron-based coagulants: optimization by response surface methodology. Chem Eng J 200–202:39–51

55. Loncnar M, Zupančič M, Bukovec P, Justin MZ (2010) Fate of saline ions in a planted landfill site with leachate recirculation. Waste Manag 30:110–118

56. Meeroff DE, Bloetscher F, Reddy DV, Gasnier F, Jain S, Mcbarnette A, Hamaguchi H (2012) Application of photochemical technologies for treatment of landfill leachate. J Hazard Mater 209–210:299–307

57. Mehmood MK, Adetutu E, Nedwell DB, Ball AS (2009) In situ microbial treatment of landfill leachate using aerated lagoons. Bioresour Technol 100:2741–2744

58. Modin H, Persson KM, Andersson A, Van Praagh M (2011) Removal of metals from landfill leachate by sorption to activated carbon, bone meal and iron fines. J Hazard Mater 189:749–754

59. Mohajeri S, Aziz HA, Isa MH, Bashir MJK, Mohajeri L, Adlan MN (2010) Influence of Fenton reagent oxidation on mineralization and decolorization of municipal landfill leachate. J Environ Sci Health A 45:692–698

60. Mohan S, Gandhimathi R (2009) Removal of heavy metal ions from municipal solid waste leachate using coal fly ash as an adsorbent. J Hazard Mater 169:351–359

61. Oti D, Thomas K, Omisca E, Howard J, Trotz M (2011) Adsorption of arsenic onto Kemiron in a landfill leachate. Toxicol Environ Chem 94:239–251

62. Palaniandy P, Adlan MN, Aziz HA, Murshed MF (2010) Application of dissolved air flotation (DAF) in semi-aerobic leachate treatment. Chem Eng J 157:316–322

63. Panizza M, Delucchi M, Sirés I (2010) Electrochemical process for the treatment of landfill leachate. J Appl Electrochem 40:1721–1727

64. Peng Y, Zhang S, Zeng W, Zheng S, Mino T, Satoh H (2008) Organic removal by denitritation and methanogenesis and nitrogen removal by nitritation from landfill leachate. Water Res 42:883–892

65. Poznyak T, Bautista GL, Chaírez I, Córdova RI, Ríos LE (2008) Decomposition of toxic pollutants in landfill leachate by ozone after coagulation treatment. J Hazard Mater 152:1108–1114

66. Renou S, Givaudan J, Poulain S, Dirassouyan, Moulin P (2008) Landfill leachate treatment: review and opportunity. J Hazard Mater 150:468–493

67. Rocha EMR, Vilar VJP, Fonseca A, Saraiva I, Boaventura RAR (2011) Landfill leachate treatment by solar-driven AOPs. Sol Energy 85:46–56

68. Saetang J, Babel S (2012) Biodegradation of organics in landfill leachate by immobilized white rot fungi, Trametes versicolor BCC 8725. Environ Technol 1–10

69. Salem Z, Hamouri K, Djemaa R, Allia K (2008) Evaluation of landfill leachate pollution and treatment. Desalination 220:108–114

70. Shou-Liang H, Bei-Dou X, Hai-Chan Y, Shi-Lei F, Jing S, Hong-Liang L (2008) In situ simultaneous organics and nitrogen removal from recycled landfill leachate using an anaerobic–aerobic process. Bioresour Technol 99:6456–6463

71. Singh SK, Townsend TG, Mazyck D, Boyer TH (2012) Equilibrium and intraparticle diffusion of stabilized landfill leachate onto micro- and meso-porous activated carbon. Water Res 46:491–499

72. Spagni A, Marsili-Libelli S (2009) Nitrogen removal via nitrite in a sequencing batch reactor treating sanitary landfill leachate. Bioresour Technol 100:609–614
73. Sun J, Li X, Feng J, Tian X (2009) Oxone/Co2+ oxidation as an advanced oxidation process: comparison with traditional Fenton oxidation for treatment of landfill leachate. Water Res 43:4363–4369
74. Syafalni, Lim HK, Ismail N, Abustan I, Murshed MF, Ahmad A (2012) Treatment of landfill leachate by using lateritic soil as a natural coagulant. J Environ Manag 112:353–359
75. Tsilogeorgis J, Zouboulis A, Samaras P, Zamboulis D (2008) Application of a membrane sequencing batch reactor for landfill leachate treatment. Desalination 221:483–493
76. Turro E, Giannis A, Cossu R, Gidarakos E, Mantzavinos D, Katsaounis A (2011) Electrochemical oxidation of stabilized landfill leachate on DSA electrodes. J Hazard Mater 190:460–465
77. Tzoupanos ND, Zouboulis AI, Zhao YC (2008) The application of novel coagulant reagent (polyaluminium silicate chloride) for the post-treatment of landfill leachates. Chemosphere 73:729–736
78. Umar M, Aziz H, Yusoff MS (2010) Trends in the use of Fenton, electro-Fenton and photo-Fenton for the treatment of landfill leachate. Waste Manag 30:2113–2121
79. Umar M, Aziz HA, Yusoff MS (2011) Assessing the chlorine disinfection of landfill leachate and optimization by response surface methodology (RSM). Desalination 274:278–283
80. Vedrenne M, Vasquez-Medrano R, Prato-Garcia D, Frontana-Uribe BA, Ibanez JG (2012) Characterization and detoxification of a mature landfill leachate using a combined coagulation–flocculation/photo Fenton treatment. J Hazard Mater 205–206:208–215
81. Wang Y, Li X, Zhen L, Zhang H, Zhang Y, Wang C (2012) Electro-Fenton treatment of concentrates generated in nanofiltration of biologically pretreated landfill leachate. J Hazard Mater 229–230:115–121
82. Wiszniowski J, Robert D, Surmacz-Gorska J, Miksch K, Weber JV (2006) Landfill leachate treatment methods: a review. Environ Chem Lett 4:51–61
83. Xiu-Fen L, Barnes D, Jian C (2011) Performance of struvite precipitation during pretreatment of raw landfill leachate and its biological validation. Environ Chem Lett 9:71–75
84. Xu ZY, Zeng GM, Yang ZH, Xiao Y, Cao M, Sun HS, Ji LL, Chen Y (2010) Biological treatment of landfill leachate with the integration of partial nitrification, anaerobic ammonium oxidation and heterotrophic denitrification. Bioresour Technol 101:79–86
85. Xu XY, Zeng GM, Peng YR, Zeng Z (2012) Potassium persulfate promoted catalytic wet oxidation of fulvic acid as a model organic compound in landfill leachate with activated carbon. Chem Eng J 200–202:25–31
86. Yahmed AB, Saidi N, Trabelsi I, Murano F, Dhaifallah T, Bousselmi L, Ghrabi A (2009) Microbial characterization during aerobic biological treatment of landfill leachate (Tunisia). Desalination 246:378–388

87. Yang Z, Zhou S (2008) The biological treatment of landfill leachate using a simultaneous aerobic and anaerobic (SAA) bio-reactor system. Chemosphere 72:1751–1756
88. Ye Z, Yu H, Wen L, Ni J (2008) Treatment of landfill leachate by immobilized microorganisms. Sci China Ser B 51:1014–1020
89. Yimin S, Qingbao G, Tichang S, Fasheng L, Yiting P (2008) Analysis and removal of organic pollutants in biologically treated landfill leachate by an inorganic flocculent composite of Al(III)–Mg(II). Ann N Y Acad Sci 1140:412–419
90. Ying D, Peng J, Xu X, Li K, Wang Y, Jia J (2012a) Treatment of mature landfill leachate by internal micro-electrolysis integrated with coagulation: a comparative study on a novel sequencing batch reactor based on zero valent iron. J Hazard Mater 229–230:426–433
91. Ying D, Xu X, Li K, Wang Y, Jia J (2012b) Design of a novel sequencing batch internal micro-electrolysis reactor for treating mature landfill leachate. Chem Eng Res Des 90:2278–2286
92. Yue X, Li XM, Wang DB, Shen TT, Liu X, Yang Q, Zeng GM, Liao DX (2011) Simultaneous phosphate and CODcr removals for landfill leachate using modified honeycomb cinders as an adsorbent. J Hazard Mater 190:553–558
93. Zhang H, Choi HJ, Canazo P, Huang CP (2009a) Multivariate approach to the Fenton process for the treatment of landfill leachate. J Hazard Mater 161:1306–1312
94. Zhang T, Ding L, Ren H (2009b) Pretreatment of ammonium removal from landfill leachate by chemical precipitation. J Hazard Mater 166:911–915
95. Zhang H, Ran X, Wu X, Zhang D (2011) Evaluation of electro-oxidation of biologically treated landfill leachate using response surface methodology. J Hazard Mater 188:261–268
96. Zhao X, Qu J, Liu H, Wang C, Xiao S, Liu R, Liu P, Lan H, Hu C (2010) Photoelectrochemical treatment of landfill leachate in a continuous flow reactor. Bioresour Technol 101:865–869

CHAPTER 2

Sustainability of Domestic Sewage Sludge Disposal

CLAUDIA BRUNA RIZZARDINI AND DANIELE GOI

2.1 INTRODUCTION

Large-scale cropland application of municipal wastewater was first practiced about 150 years ago after flush toilets and sewer systems were introduced into cities in Western Europe and North America. Wastewater was discharged without any treatment and receiving watercourses became heavily polluted [1]. Many "sewage farms" were designated as a preferred alternative to the direct discharge of raw sewage into waterways [2]; in 1875, this "land treatment" served England and major cities in Europe and with the new century it has also started in the United States. Sewage farms played a role in decreasing pollution in the receiving streams, and also creating several environmental sanitation problems: hydraulic and pollutant land overloading caused clogging of soil pores, soil water logging, odors and contamination of food crops [1]. This procedure was

---

© 2014 by the authors; licensee MDPI, Basel, Switzerland. Sustainability 2014, 6(5), 2424-2434; doi:10.3390/su6052424. Creative Commons Attribution license (http://creativecommons.org/licenses/by/3.0/). Used with authors' permission.

gradually abandoned with the development of more effective technologies and building municipal sewage treatment systems. Land application of sewage sludge starts with the treatment of municipal wastewater and the consequent production of an end product, a solid waste, consisting in a concentrated suspension of solids high in organics and biodegradable compounds. One of the most widely used process for the treatment of wastewaters from medium to large populations that has found application in almost all of the countries of the world is the activated sludge process. It is a biological treatment which uses a mass of microorganisms to aerobically treat wastewater. It is widely accepted that the original process is attributed to the experimental work undertaken by Dr. Edward Ardern and Mr. William Lockett and carried out at the Davyhulme Sewage Works, which at that time were operated by the Manchester Corporation, with the cooperation of Dr. Fowler in 1914 [3]. The process was developed for the treatment of domestic wastewater and it has since been adapted for removing biodegradable organics from industrial wastewaters [4]. Owing to the physical-chemical processes involved in its treatment, sewage sludge tends to concentrate heavy metals and poorly biodegradable trace organic compounds, as well as potentially pathogenic organisms (viruses, bacteria, etc.) presented in wastewaters [5]. Since the late 1970s, source control and industrial wastewater pretreatment programs were applied to limit the discharge of industrial constituents into municipal sewers resulting in a consistent reduction of trace elements in wastewater and sewage sludge. Over the past 30 years and until now, many studies were started to understand and predict the toxicity and the fate of toxic substances and pathogens in sewage sludge when they are applied to soils [1]. Results constituted a benchmark for the development of guidelines in the United States and in western European countries. For more than 20 years, the sewage sludge directive 86/278/EEC has encouraged the use of sewage sludge in agriculture, suggesting at the same time regulation of its use to prevent possible harmful effects on soil, vegetation, animals and humans. The key concept of the directive is to consider sewage sludge as a valuable resource: in fact it is rich in plant macro and micronutrients and its application can, in the long term, improve soil fertility [6]. In Europe the most pragmatic and environmentally sustainable approach to manage sludge from wastewater treatment plants is actually recycling it on agricultural land [7]. This strategy is supported by many scientific and regulatory authorities even if it is not adopted by all European countries. In fact, the public debate on the use of sludge in agriculture is well cared about in some Member States: Northern Europe Netherlands and Flanders have prevented almost all use of sewage sludge in agriculture. Other countries such as Denmark, Germany and Sweden have established new regulations believed sufficiently stricter to reduce risks, but the political discussion is open yet. Until now, in Italy, Greece and Spain this argument is not so known, perhaps due to information deficiency [8].

In Italy, land spreading manages about the 32% of the total volume of sewage sludge produced, but the regional situation is diversified: only a few Italian regions apply to soil large amounts of sewage sludge and sometimes new advanced local regulations about land spreading are applied. As a consequence, landfilling is used as a forced choice for sludge not in accordance with restricted limits; this way to dispose sewage sludge represents the only management strategy in several Regions. Composting and anaerobic digestion are another ways to get to a treated product for agriculture use, but these practices are second order management solutions together with incineration and the production line of concrete, asphalt and bricks.

Apart from political choices and management strategies the matter is to establish the sustainability of sewage sludge agricultural disposal evaluating the level of pollution of both sewage sludge and amended soils and the safety of its application. The last official document from the European Union about sewage sludge should be a useful instrument to get a sense about the level of pollution of sewage sludge and soils: about twenty years passed from the first and the last European regulation on sludge in agricultural use (86/278/EEC) [9]. In this final directive, only heavy metals as potential pollutants in sewage sludge were considered, even if many studies have evidenced potential health hazards associated with sludge-borne toxic organic [10,11]. In the meantime, several States found a high concentration of several contaminants and set out limit values of concentration (e.g., Denmark with Linear Alchil Benzen Sulfonate); as a consequence, in 2000, the European Community elaborated a draft (Working document on sewage sludge [12]) suggesting new limits for some toxic organics on the basis of precautionary principles. The European Commission is currently assessing and reviewing the sewage sludge Directive and it has launched a study to gather existing information on the environmental, economic, and social, as well as the health impacts of sewage sludge use on land [6]. As a matter of fact, agreement on regulating organic contaminants in sewage sludge is likely to emerge as one of the most controversial aspects of the consultation on the revised Sludge Directive [7].

This paper evaluates the level of pollution of sewage sludge and amended agricultural soils applying the dictates of the Third Working document on sludge. Medium-small size wastewater treatment plants are widespread in Italy and sludge is often applied in soils for agricultural use in the nearby areas. The Italian Environmental Agency controls quality and collects data on sludge but not many studies are reported about the system sludge-amended terrains. This work makes little contribution in this field, considering the particular situation of small dimension of domestic and urban plants in a not heavily industrialized area. Analysis were performed on three typical and different sludge from wastewater treatment plants of small Italian municipalities; soils receiving continuous applications of sewage sludge produced by these plants were also investigated.

2.2 EXPERIMENTAL SECTION

This case study reports data from a wastewater-sludge management organization for small communities located in the north-east of Italy: the Poiana Waterworks Ltd. manages a water-wastewater integrated system of various small municipalities and one little industrial district in an area characterized by common rural economy. The society is in charge of the whole water management, from collecting to treatment of mainly domestic and urban wastewaters, in 31 small treatment plants (from 2.500 to 6.000 inhabitant equivalents).

This study deals with sewage sludge produced by three different wastewater treatment plants (Table 1), selected according to population served and type of treated wastewaters in compliance with Directive 91/271/EEC [13]: the first plant treats typical domestic wastewaters with a 4.000 inhabitant equivalent (i.e.) potential; the second (5.000 i.e.) receives domestic wastewater with agro-industrial liquid waste contributes, typically milk

Sustainability of Domestic Sewage Sludge Disposal

Table 1. Concentration and limit values of heavy metals in sewage sludge produced in small communities (mg kg^{-1} DM).

Sample	Cd	Cr	Cu	Ni	Pb	Zn	Hg
Domestic	1.357 ± 0.026	39.58 ± 2.96	456.6 ± 7.3	31.21 ± 1.13	70.69 ± 1.46	1260.8 ± 18.1	0.580 ± 0.086
Urban with agro-industrial intrusions	0.331 ± 0.039	45.10 ± 0.29	455.2 ± 2.1	21.55 ± 0.46	18.70 ± 0.57	286.3 ± 4.7	1.320 ± 0.210
Urban with industrial intrusions	1.099 ± 0.032	57.16 ± 1.16	727.2 ± 7.3	34.09 ± 0.94	57.84 ± 0.49	510.5 ± 3.8	1.730 ± 0.096
Limit 2000	10	1000	1000	300	750	2500	10
Limit 2015	5	800	800	200	500	2000	5
Limit 2025	2	600	600	100	200	1500	2

working and little dairy production. The last plant (6.000 i.e.) collects urban wastewater including liquid wastes from a service station with car wash facilities.

The paper considers the level of pollution of sewage sludge following evaluation criteria and methods of the working document on sludge. Analysis of heavy metals (Cd, Cr, Cu, Ni, Pb, Zn and Hg) and organic compounds (PCB, PCDD/F, PAH, LAS, NPE, DEHP and EOX) were performed on these three sewage sludge samples during 2009. Detailed information about methods of analysis was published in a previous article by Rizzardini et al. (2012) [6]. These data were compared with limits for agricultural use defined by the working document on sludge.

In the study several soils interested by application of these three types of sewage sludge were also considered: on the base of sludge quantities applied to each single plot in a time lapse of ten years and annual chemical analyses of sewage sludge applied, average annual quantities of heavy metals introduced into each sample was calculated; finally, data were related to limits for soil application suggested by the working document on sludge.

From a group of 24 samples drawn from the amended plots that received great inputs of sewage sludge from the selected wastewater treatment plants, three samples were chosen.

Over a period of ten years, the first plot had received 14.53 t/ha, the second 8.64 t/ha and the third 12.45 t/ha of a mixture of the three sewage sludge. Sludge was applied to soil following traditional agronomic rates mixing it with first 15 cm top soil surface.

2.3 RESULTS AND DISCUSSION

2.3.1 SEWAGE SLUDGE POLLUTION

The Working document on sludge identified two main types of sewage sludge source of pollution: heavy metals and some groups of organic compounds. For these contaminants several concentration limits were suggested, also considering future medium (2015) and long term (2025) proposed limits.

In Table 1 average concentrations of toxic metals found in the three sewage sludge analyzed are reported; it can be noted all data are widely under limits proposed by Working document on sludge.

Generally, Cd, Pb and Zn should present biggest concentrations in sewage sludge from domestic wastewaters that can be a source of pollution for soils. Feces contribute 60%–70% of the load of Cd, Zn, Cu and Ni in domestic wastewater and >20% of the input of these elements in mixed wastewater from domestic and industrial premises. The other principal sources of metals in domestic wastewater are body care products, pharmaceuticals, cleaning products and liquid wastes [14]. Based on mass flow studies, the runoff from roofs and streets contribute 50%–80% of Cd, Cu, Pb and Zn to the total mass flow in domestic sludge [15]. Data collected in this study show that all metals considered in the Working document have similar rate of presence even if at concentrations far from limits.

Nevertheless, some other peculiarities of the case study can be underlined. Sewage sludge of sample nr. 2 has metal charges lower than others, even if it was originated from urban wastewater. These results are reasonable if we consider that the plant receives and treats mainly wastewaters

Sustainability of Domestic Sewage Sludge Disposal

from an agro-industrial pole. Sample nr. 3 has the highest concentrations of Cr, Ni, Cu and Hg: these values seem realistic if we consider that this plant treats urban wastewaters with intrusions from a service station with car wash facilities. Cr contamination is probably due to wear of tires and road pavements, which are the two major emission sources identified; for Ni, in general the traffic sector dominates the emissions, even if there are reports about Ni release from older concrete surfaces [16]. High load of Cu and Hg are probably justified by the fact that they are widespread contaminants: in fact, plumbing is the main source of Cu in hard water areas, contributing >50% of the Cu load [17]. Instead Hg is an ubiquitous pollutants and its stock could be dominated by the use of amalgam or by its deposits in the wastewater system [18].

Besides heavy metals, the working document on sludge regulated several persistent organic pollutants (POPs): in fact, as an end product of sewage treatment, sludge accumulates many substances, especially lipophilic ones, which are present in the wastewater but not fully degraded during wastewater treatment. The spectrum of organic substances of anthropogenic origin occurring in sewage sludge is extremely wide and constantly changing [18]. Drescher-Kaden et al. [19] reported that 332 organic substances, with the potential to exert a health or environmental hazard, had been identified in German sludge and 42 of these were regularly detected in sludge [7]. Working document on sludge proposed limit values for seven main groups of POPs, taking example from past legislation of several European Member States. The groups of compounds whose regulation has been suggested are: absorbable organic halogens (AOX), polychlorinated biphenyls (PCB), dibenzo-p-dioxins and furans (PCDD/F), polycyclic aromatic hydrocarbons (PAH), di-2-(ethyl-hexyl) phthalate (DEHP), non-ylphenol and nonylphenolethoxylates (NPE) and linear alkylbenzene sulfonates (LAS). In this study we quantified organic halogen pollution with extractable organic halogen (EOX) parameter instead of AOX because of its suitability to characterized complex two-phase matrices as sludge. In Table 2 results of the chemical analysis of organic contaminants were presented: it can be noted a general respect of regulation and each sewage sludge has concentrations much lower than limits. No very significant differences among samples were noted according to potentiality, origin of wastewaters and types of sludge technologies, with the exception of LAS content in sample nr. 3. The plant related to this sludge treated wastewaters from a service station with car wash facilities and LAS are anionic surfactants employed in detergents and cleaning products. EOX values follow results obtained in another study by Goi et al. [20]: although research has been conducted in areas of the same district about 4 years ago with respect to the analysis of this study, our results present the same order of magnitude. Most of the organics included in this parameter have been classified among the most dangerous pollutants in the lists of many environmental protection agencies of all over the world [20]. Measure of EOX content in sludge could give a good estimation of the level of harm due to the presence of halorganics in this type of waste and EOX can be used as a measure parameter for sewage sludge quality. EOX comprises contaminants as PCBs, PCDDs and PCDFs: PCBs concentrations are lower than the detection limit and it is probably due to the fact that many countries and intergovernmental organizations have now banned or severely restricted the production, use, handling, transport and disposal of these contaminants. PCDDs and PCDFs are ubiquitous pollutants not commercially produced but formed as trace amounts of undesired impurities in the manufacture of other chemicals, such as chlorinated phenols and their derivatives, chlorinated diphenyl ethers and PCBs: a variety of sources were identified, including sewage sludge [21] and garden composts [22], in which they can be formed naturally. Although PAHs are ubiquitous chemical contaminants, toxic to humans and potentially threatening chronic long-term: compared to PCDDs/PCDFs they derive from anthropogenic and natural activities. The concentrations of PAHs are really lower than those found in a conventional wastewater treatment plant (WWTP) located in northern Italy with a capacity of about 400.000 i.e., which receives mixing wastewater and presents a concentration of PAHs of about 3 mg kg^{-1} DM [23]. Finally, NPEs and DEHP are used in anthropogenic activities and their concentration are much lower than limit values of concentration: NPEs are used as surface active agents in cleaning products, cosmetics and hygienic products, in emulsifications of paints and pesticides and are slowly being phased-out of the market; DEHP is the most common of the phthalate esters, which are widely used industrial chemicals, serving as important additives, which produce flexibility in polyvinylchloride (PVC) resins [24].

It must be underlined that limit values for POPs were established on precautionary concentrations and were not based on ecotoxicological

Sustainability of Domestic Sewage Sludge Disposal

Table 2. Concentration and limit values of organic pollutants in sewage sludge from small communities (mg kg^{-1} DM).

Sample	EOX (AOX)	ΣLAS	DEHP	ΣNPE	ΣPAH	ΣPCB	ΣPCDD/ PCDF *
Domestic	9.03	48.8	6.20	1.57	1.15	<0.0035	49
Urban with agro-industrial intrusions	9.79	64.5	13.04	1.01	0.55	<0.0035	41
Urban with industrial intrusions	10.10	239.1	6.31	0.80	1.03	<0.0035	40
Limit 2000	500	2600	100	50	6	0.8	100

* **values** expressed as ng TE kg^{-1} DM.

researches. In particular, the limit values for AOX, PCBs and PCDD/Fs are intended as precautional and are not justified solely by toxicological implications [25]. Moreover the EU proposals include limit values for the sum of 9 PAHs, but it is not clear what criteria have been used to select these compounds. Subsequently, studies have been done to evaluate the relevance of organic micro-pollutants in sewage sludge and to suggest guide values taking into account different situations of European countries [26]. Actually, the question remains open, and the European Commission has launched a study to evaluate the presence of emerging pollutants in sewage sludge which could contaminate terrestrial and aquatic environment when sludge is used in agriculture [27].

2.3.2 SOIL POLLUTION

Working document on sludge took in consideration also the possible contamination of soil following sewage sludge application. Because of their sparingly soluble nature and their limited uptake by plants, heavy metals tend to accumulate in the surface soil and become part of the soil matrix. With repeated applications of sewage sludge, heavy metals could

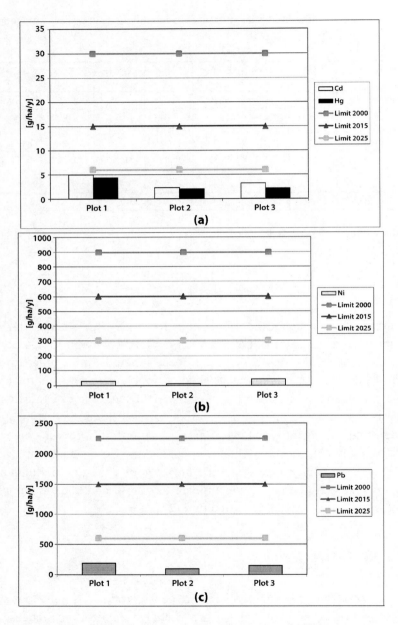

Figure 1. Effective amounts of heavy metals added annually to soils of the wastewater district and comparison with limit values suggested by the working document on sludge. (a) Cd and Hg; (b) Ni; (c) Pb; (d) Cu and Cr; (e) Zn.

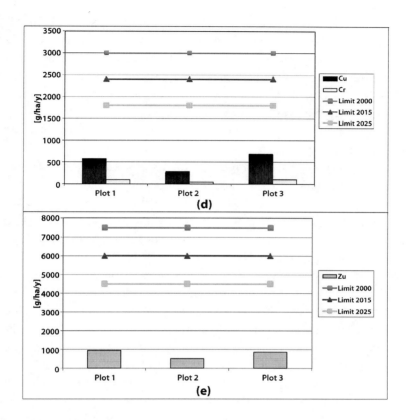

Figure 1. Continued.

accumulate to levels toxic to crops and build up to potentially harmful levels in humans, animals and wildlife that consume the crops [28]. In addition they can be detrimental to key microbial processes related to the cycling of nutrients and the turnover of organic matter [29]. The organic contaminants introduced into the soil through the agricultural utilization of sewage sludge are subjected to more rapid degradation, with the result that accumulations in this field can be effectively discounted [18]. It must be noticed that the presence of toxic metals in soil can severely inhibit the biodegradation of organic contaminants [30]. In our case, we focused on the study of the contribution of heavy metals on soils interested by land applications. Further, it has been calculated a parameter just considered in Sludge Directive, but often disregarded by national legislations: the average load of heavy metals added annually to soils expressed as gram of contaminant supplied per hectare and per year (g ha^{-1}·year^{-1}). Considering this parameter, plots of lands interested by ten years applications of the three types of sewage sludge described above have been analyzed. Results of this study are presented in Figure 1: each soil sample is in compliance with limit values suggested by the working document on sludge, both for the actual (between 2000 and 2015) and the future period (limit 2025). Limits proposed are precautionary values for the protection of long-term soil quality having a regard to background concentrations in European agricultural soils [31]. This study highlighted some criticisms: even if in accordance with the proposals of European Commission, these results show a potential trend for future exceeding of the rate concentration limit, especially for Cd. Secondly, the high content of Cu and Zn in sewage sludge suggests that the potential flow of these elements to soils is high with a significant accumulation. These components indicate a possible environmental impact in a longer time perspective.

These outcomes agree with and confirm the study conducted for the European Commission by Millieu/WRc/RPA, which suggests that consideration needs to be given to adjusting the maximum permissible soil metal limits in Directive 86/278/EEC for cadmium and zinc [31].

2.4 CONCLUSIONS

Since sewage sludge as the main by-product of WWTP in Italy is often recycled in agriculture, it can be a guide parameter to observe regional environmental pollution from wastewater discharge. Analyses coming from this case study confirm that sewage sludge from small and medium communities have a good quality and compliance with the criteria of the Working document on sludge ensuring a safe application. However continuous and repeated applications may cause possible future heavy metals enrichment of the soil system. Data reported highlight the need for

Sustainability of Domestic Sewage Sludge Disposal

further research programs to quantify the real level of pollution of soils interested by long period land application of sewage sludge. It is also a fact that sources of heavy metals soil contamination can be found not only in sewage sludge application, but also in other agricultural practices (e.g., application of pesticides, fertilizers, animal manure). Research programs should consider and evaluate application sites (pH, CEC, buffering capacity, organic matter and clay content) and management of plots. The case study of Poiana Waterworks Ltd. represents a little but unique project in which the dictates of the working document on sludge are completely applied. This method should be used for valuable support in regional government regulations about sewage sludge application in agriculture following European suggestions and waiting for new directives and regulations.

REFERENCES

1. Water Environment Federation. Available online: http://www.wef.org/biosolids-factsheets.aspx (accessed on 1 December 2013).
2. U.S. Environmental Protection Agency. Process Design Manual. Land Application of Municipal Sludge, 1st ed.; U.S. Environmental Protection Agency Center for Environmental Research Information: Cincinnati, OH, USA, 1983; pp. 1–5.
3. Activated Sludge: Past, Present and Future. Available online: http://www.activatedsludgeconference.com (accessed on 3 December 2013).
4. Sustarsic, M. Wastewater treatment: Understanding the activated sludge process. Chem. Eng. Prog. 2009, 105, 26–29.
5. Jorge, F.C.; Dinis, M.A.P. Sewage sludge disposal with energy recovery: A review. ICE Proc.—Waste Res. Manag. 2012, 166, 14–28.
6. Rizzardini, C.B.; Contin, M.; de Nobili, M.; Goi, D. Sewage sludge quality from small wastewater treatment plants. ICE Proc.—Waste Res. Manag. 2012, 165, 67–78.
7. Smith, S.R. Organic contaminants in sewage sludge biosolids and their significance for agricultural recycling. Philos. Trans. R. Soc. A 2009, 367, 4005–4041.
8. European Commission. Disposal and Recycling Routes for Sewage Sludge. Part 1—Sludge Use Acceptance Report. Available online: http://ec.europa.eu/environment/waste/sludge/pdf/sludge_disposal1.pdf. (accessed on 14 April 2014).
9. Council Directive. Council directive on the protection of the environment, and in particular of the soil, when sewage sludge is used in agriculture. Offic. J. Eur. Comm. 1986, 181, 0006–0012.
10. O'Connor, G.A.; Haney, R.L.; Ryan, J.A. Bioavailability to plants of sludge-borne toxic organics. Rev. Environ. Contam. Toxicol. 1991, 121, 129–155.
11. Leschber, R. Part I: Evaluation of the Relevance of Organic Micro-Pollutants in Sewage Sludge. In Background Values in European Soils and Sewage Sludges;

Gawlik, B.M., Bidoglio, G., Eds.; European Commission: Brussels, Belgium, 2006.

12. European Commission. Working Document on Sludge 3rd Draft; DG Environment, European Commission: Brussels, Belgium, 2000.

13. Council Directive. Council Directive 91/271/EEC of 21st May 1991 concerning urban waste water treatment. Offic. J. Eur. Comm. 1991, 135, 40–52.

14. ICON Consultants. Pollutants in Urban Waste Water and Sewage Sludge—Final Report for DG Research; Office for Official Publications of the European Communities: Luxembourg, Luxembourg, 2001.

15. Boller, M. Tracking heavy metals reveals sustainability deficits of urban drainage systems. Water Sci. Technol. 1997, 35, 77–87.

16. Persson, D.; Kucera, V. Release of metals from buildings, constructions and products during atmospheric exposure in Stockholm. Water Air Soil Pollut. 2001, 1, 133–150.

17. Bergback, B.; Johansson, K.; Mohlander, U. Urban Metal Flows—A case study of Stockholm. Review and Conclusions. Water Air Soil Pollut. 2001, 1, 3–24.

18. Schnaak, W.; Kuchler, T.; Kujawa, M.; Henschel, K.P.; Suszenbach, D. Organic contaminants in sewage sludge and their ecotoxicological significance in the agricultural utilization of sludge. Chemosphere 1997, 35, 5–11.

19. Drescher-Kaden, U.; Bruggeman, R.; Matthes, B.; Matthies, M. Contents of organic pollutants in German sewage sludges. In Effects of Organic Contaminants in Sewage Sludge on Soil Fertility, Plants and Animals; Hall, J.E., Sauerbeck, D.R., L'Hermite, P., Eds.; Commission of the European Communities: Luxembourg, Luxembourg, 1992; pp. 14–34.

20. Goi, D.; Tubaro, F.; Dolcetti, G. Analysis of metals and EOX in sludge from municipal wastewater treatment plants: A case study. Waste Manag. 2006, 26, 167–175.

21. Öberg, L.G.; Andersson, R.; Rappe, C. De Novo Formation of Hepta- and Octachlorodibenzo-p-Dioxins from Pentachlorophenol in Municipal Sewage Sludge. In Organohalogen Compounds 9; Finnish Institute of Occupational Health: Helsinki, Finland, 1992; pp. 351–354.

22. Öberg, L.G.; Wagman, N.; Andersson, R.; Rappe, C. De Novo Formation of PCDD/Fs in Compost and Sewage Sludge—A Status Repor. In Organohalogen Compounds 11; Environmental Protection Agency: Vienna, Austria, 1993; pp. 297–302.

23. Torretta, V.; Katsoyiannis, A. Occurrence of polycyclic aromatic hydrocarbons in sludges from different stages of a wastewater treatment plant in Italy. Environ. Technol. 2013, 34, 937–943.

24. Cifci, D.I.; Kinaci, C.; Arikan, O.A. Occurrence of phthalates in sewage sludge from three wastewater treatment plants in Istanbul. Clean–Soil Air Water 2013, 41, 851–855.

25. Sauerbeck, D.R.; Leschber, R. Effects of Organic Contaminates in Sewage Sludge on Soil Fertility, Plants and Animals; Commission of the European Community: Luxembourg, Luxembourg, 1992; pp. 1–13.

26. Díaz-Cruz, M.S.; García-Galán, M.J.; Guerra, P.; Jelic, A.; Postigo, C.; Eljarrat, E.; Farré, M.; López de Alda, M.J.; Petrovic, M.; Barceló, D. Analysis of

selected emerging contaminants in sewage sludge. Trends Anal. Chem. 2009, 28, 1263–1275.

27. Environmental Protection Branch (EPB) 296. Land Application of Municipal Sewage Sludge Guidelines; Water Security Agency: Moose Jaw, SK, Canada, 2004.

28. Obbard, J.P. Ecotoxicological assessment of heavy metals in sewage sludge amended soils. Appl. Geochem. 2001, 16, 1405–1411.

29. Maslin, P.; Maier, R.M. Rhamnolipid-enhanced mineralization of phenanthrene in organic-metal co-contaminated soils. Bioremediat. J. 2000, 4, 295–308.

30. European Commission. Working Document: Sludge and Biowaste; DG Environment, European Commission: Brussels, Belgium, 2010.

31. European Compost Network (ECN). Comments to EC DG Env. "Working Document on Sludge and Biowaste". Available online: http://circa.europa.eu/Public/irc/env/rev_sewage/library?l=/comments_document_2010/other_stakeholders/bio-sludge-working-doc/_EN_1.0_&a=d (accessed on 14 April 2014).

PART II

SEWAGE SLUDGE TREATMENTS

CHAPTER 3

Composting Used as a Low Cost Method for Pathogen Elimination in Sewage Sludge in Mérida, Mexico

DULCE DIANA CABAÑAS-VARGAS,
EMILIO. DE LOS RÍOS IBARRA, JUAN. P. MENA-SALAS,
DIANA Y. ESCALANTE-RÉNDIZ, AND RAFAEL ROJAS-HERRERA

3.1 INTRODUCTION

The present work was developed in the city of Mérida, located on the Yucatán Península with 900,000 inhabitants approximately including its surrounding areas [1]. All area is formed by a very permeable karst surface [2]. Due to karstic conditions, there are a lot of fractures in the subsoil, the aquifer receives water from rainfall including any pollution that is picked up from the land surface [2,3]. The aquifer is also the unique source of water in the region. The main sources of pollution of the aquifer are the wastewater and sludge from the municipal wastewater system (septic tanks, small treatment plants and waste water from small industry).

The main goal of this study was to test the compost process as an economical and efficient method for pathogen elimination in sludge from a

---

© 2013 by the authors; licensee MDPI, Basel, Switzerland. Sustainability 2013, 5(7), 3150-3158; doi:10.3390/su5073150. Creative Commons Attribution license (http://creativecommons.org/licenses/by/3.0/).

municipal waste water system. In the case of the city of Merida, Mexico, this process turns out to be important because at present the sludge from the municipal wastewater system does not receive any treatment before being deposited on soils, so it might represent a potential source to spread pathogens.

The composting process has shown to be an economic method, capable of eliminating or reducing the pathogenic microorganisms present in the material being treated by composting.

According to Epstein E. [4], composting is very effective in destroying pathogens, as a result of temperature—time relationship.

3.1.1 BACKGROUND

Most houses in Merida have septic tanks as a sewage system. Once the septic tank is full, its content is collected and discharged into an oxidation pond system where sludge and water are separated by decantation. Separated sludge is discharged on the ground where it is sun dried. Endemic birds and insects like flies and worms have continuous contact with this waste. For flies, it becomes an ideal place for laying its eggs because of the amount of nutrients it has. As a consequence, they become a way of transporting pathogenic microorganisms with the potential of spreading human diseases.

As expected, municipal solid waste, sludge from municipal waste water (biosolids), food waste and yard waste, contain human pathogenic organisms.

Figure 1 illustrates an oxidation pond constructed in the city of Merida as a Municipal Waste Water System which is the only treatment available for wastewater in the area, for both household and micro-industry wastewater. The plant was designed to provide treatment to 9550 m^3 of wastewater per month; however, it receives between 10,000 and 12,000 m^3 [5].

Composting, if carried out properly, is an effective process to destroy contamination indicators and pathogens like faecal coliforms and Salmonella pp as a result of the relationship between time spent at elevated temperatures and pathogen destruction. For example, if the compost reaches a temperature of 60 °C during 30 min or 70 °C during 4 min, Salmonella

Composting Used as a Low Cost Method for Pathogen Elimination

Figure 1. Oxidation ponds system used in Mérida, Mexico as a Municipal Waste Water System.

in biosolids can be destroyed on the other hand total coliforms and faecal coliforms show great reduction when temperatures exceed 55 °C to 65 °C [4].

During the thermophilic stage of composting, an increase of temperature can occur through the sludge mass as a result of the microorganisms' activity. Thus pathogens content within the sludge could be destroyed. Temperatures above 45 °C characterizes this period of composting [4,6,7]. During the next stage temperature descends, causing a decrease of degradation speed. At the end of the composting process, resulting material looks similar to humus, and is free of easy available organic matter, weed seeds and pathogens.

Composting is an aerobic process, thus oxygen is essential element for the microbial activity. Sludge from waste water treatments usually contains large amounts of moisture which inhibits the free flow of air. Wood or bark chips, sawdust, yard trimmings and pruning trees have been tested like bulking agents providing free spaces for an adequate oxygenation [8,9]. For the current work yard trimmings were used. Composting by windrow has shown a good performance for sludge mixed with yard trimmings. According to Velazquez et al. [10], with this process, fecal coliforms were eliminated by 90% after a month.

The windrow composting method needs a lot of space and some mechanism for agitating the material (adding oxygen). For the city of Merida, the land is very cheap and oxygen can be added by manual turning of the pile. Therefore, composting can be achieved with only a small input of resources.

3.1.2 RESEARCH OBJECTIVES

- To test the windrow composting as a low cost process to eliminate pathogens contained in sewage sludge from municipal wastewater treatment.
- To observe the behavior of helminth eggs, salmonella and faecal coliforms during the windrow composting process.
- To obtain a final product that can be deposited on the soil safely in terms of environment and human health.

3.2 METHODOLOGY

3.2.1 COMPOSTING PROCESS

Composting is not a common practice in the region; the present work was developed as a pilot experiment.

Two real size-composting piles were built for this work over a concrete slab nearby the municipal waste water system in Mérida, Mexico. In this city, the municipal wastewater system is an oxidation pond (See Figure 2). Each pile was constructed by a mixture of sludge from the municipal waste water system (40%) and green waste or yard trimmings (60%) which altogether added to 1000 L. Composting was made by windrow system, turning piles manually twice a week and using Temperature, Moisture Content and pH as control parameters. The temperature was measured everyday in 9 different points inside the compost matrix (30 cm depth) using a digital thermopar Tegam 871; the samples for the other parameters were taken after turning the piles. The pH was measured using a pH meter Thermo Orion with a range between 2 to 14 and an accuracy of ±2. Composting piles were monitored during four weeks for the active stage by analyzing, twice a week the concentration of Helminth eggs, Salmonella and faecal Coliforms. After the active stage, the composting material was tested twice a month during the following two months for the same pathogens and checking the germination index.

3.2.2 LABORATORY TECHNIQUES

3.2.2.1 SALMONELLA AND FAECAL COLIFORMS

For salmonella and faecal coliforms tests, the procedures described by the NOM-004-SEMARNAT were used [11]. This methodology is based on the procedures described in the "Standard Methods for the examination of Water and Wastewater", where the techniques of Most Probable Number (MPN) were used. These procedures have been established by the Mexican environmental legislation specifically to indicate the possible presence of

Figure 2. Composting piles at real scale using windrow system and manual agitation.

Composting Used as a Low Cost Method for Pathogen Elimination 55

pathogens in sludge and also to establish the maximum permissible limits for these indicators in sludge to be deposited in soils safely.

3.2.2.2 HELMINTH EGGS

For determining helminth eggs concentration, the method described by Moodley et al. [12] was used. The method is based on three fundamental processes: washing, filtering one or more times, and then floating and sedimenting of the retrieved parasites. A flotation step is used for the isolation of helminth ova using density gradient centrifugation and a chemical solution that is saturated at a specific gravity of 1.3 so that all helminth ova having relative densities that range from 1.13 (e.g., Ascaris) to 1.27 (e.g., Taenia) are able to float in that solution. The deposit is transferred to one or more microscope slides and examined under the microscope to enumerate each species of Helminth ova using the 10x objective and the 40x objective to confirm any uncertainties.

3.2.2.3 GERMINATION INDEX

Germination index was performed by a phytotoxicity bioassay, using a liquid extract of compost for watering radish seeds (instead of cress seeds), as described by the Methodology used by the Lab Staff at University of Leeds, UK [13] which is an adaptation of the method of Zucconi, et al. [14]. 30 g of screened sample were shaken with 300 mL of distilled water for 1 h. The suspension was centrifuged for 15 min at 3000 rpm and the supernatant was filtered through a Whatman No 6 filter paper. 10 radish seeds were placed on filter paper Whatman No 1 in a Petri dish of 10cm diameter. Two ml of the extract were added to the Petri dish. Two ml of distilled water were used for control. The test was run in triplicate. Petri dishes were left on laboratory bench and after 48 h, the total length of each cress root was measured. If the seeds did not germinate, their root length was considered to be 0 mm.

Results were calculated based on the following formula:

$$GI = (\text{Total Length of roots in test place} / \text{Total Length of roots in control plate}) \times 100$$

3.2.2.4 CONTROL PARAMETERS

Values of potential of Hydrogen (pH) and Moisture Content were obtained according with the "Standard Laboratory Procedures for the Analysis of Compost" from Leeds University, UK [10,13]. and Federal Compost Quality Assurance Organization Manual, Germany [15].

3.3 RESULTS AND DISCUSSION

After 30 days of composting sludge from municipal waste water system, Salmonella was eliminated by 99%, Faecal coliforms by 96% and Helminth Eggs by 81% (See Table 1).

The potential risk of the presence of faecal coliforms in sewage sludge is because they are extremely resistant to certain conditions and their persistence for long periods of time. Therefore, faecal coliforms are used as indicators of the effectiveness of treatment processes in the destruction of bacteria, and regulate the quality of sewage sludge that can be used safely. They are also indicative of the concentration of Salmonella spp., bacteria that are usually associated with gastrointestinal diseases in humans and thus reducing faecal coliforms ideally reflecting a decrease in *Salmonella* spp. [16].

Although early in the process the sludge have had high levels of salmonella (higher than faecal coliforms), with increasing temperature during the composting process, the content of salmonella decreased to values lower than fecal coliforms. This experimental result is consistent with the nature of coliform group, which have higher rate of temperature–survival times than that of enteric pathogens like salmonella [4].

Table 1. Results including pathogens reduction in time and control parameter behaviour.

Composting Time (days)	SalmonellaMPN/ g sludge	Faecal ColMPN / g sludge	Helminth Eggs / g sludge	Control parameters		
				T °C	MC%	pH
0	2.4×10^8	2.4×10^6	146	27	52.77	7.27
4	4.6×10^7	1.1×10^6	131	56	68.57	7.7
9	2.1×10^6	4.6×10^5	87	63.5	58.61	7.87
14	1.5×10^6	2.4×10^5	38	63.0	67.23	7.73
18	7.5×10^5	2.1×10^5	32	53.6	56.21	7.71
23	3.9×10^4	1.5×10^5	29	49.1	50.14	7.69
28	1.1×10^4	1.2×10^5	29	37.5	48.86	7.67
31	3.0×10^3	9.3×10^4	27	37.2	46.72	7.64
Max. Permissible limits for sludge C class* [11]	3.0×10^2	2.0×10^6	35			

* Sludge "C" class are allowed for forest, soil improvement and agricultural use; All values are the average of 3 replicates.

The concentration of Salmonella was decreased up to by 99%, but failed to reach the maximum levels of 300 NMP/g reported by the Mexican regulations for classification in Class C (For forest, soil improvement and agricultural uses) [11]. This may be due to the nature of the sludge which comes from domestic septic tanks and consequently contains high levels of faecal contamination. Faecal coliform levels were below 2,000,000 MPN/g corresponding to the class C according to the same Mexican regulation as above [11]. This indicates that the sludge can be disposed of safely in regard to fecal coliforms.

Helminth eggs, particularly Ascari, are very resistant to changes in environmental factors, which are stressful for other microorganisms, thus present an additional challenge for sludge treatment processes. The number of helminth eggs found after the composting process is a Class C according to Mexican regulation [11]. Several authors suggest that only the processe applying high temperatures (about 50 °C or more), reduce the density and the viability of different Helminth eggs [4,6,17], this coincides with the results, since the temperatures reached in the process of composting were higher at 60 °C.

For the germination index was only possible to sample every month, the samples were performed in duplicate and showed positive results (Table 2). The first month the germination rate was 65% on average, which is consistent with results reported by other researchers [10,18,19,20]. The last test performed strong in the third month and reported a GI = 160%. According to Manser and Keeling [14], values above 100% are indicative of positive influence of compost to the germination process.

However, these values as unique criteria are not enough to make a real assessment and it is important to clarify that this procedure is only focused in the initial germination phase and it cannot predict the behavior during the whole plant growth.

3.4 CONCLUSIONS

- After 30 days of composting sludge from municipal waste water system, Salmonella was eliminated by 99%, Faecal coliforms by 96% and Helminth eggs by 81%.

Composting Used as a Low Cost Method for Pathogen Elimination 59

Table 2. Germination index with time.

Composting time (months)	Germination Index (%) **
1	65
2	98
3	160

** All values are the average of 3 replicates.

- The fecal coliform concentration decreased to levels below the maximum permissible limits
- The sludge can be disposed directly on the ground safely with respect to the concentration of fecal coliforms.
- Longer composting times and better control process are required for the elimination of Salmonella to achieve maximum permissible.
- Germination Index values indicate that the compost did not show any phytotoxicity to seeds, but the use as a soil amendment should be further tested.
- The results showed that a composting process is an efficient method for pathogens elimination in sewage sludge from municipal waste water.

3.5 RECOMMENDATIONS

- For future work, use longer composting times and a better control parameters
- For Merida city, it will be very useful to use composting for sanitazion of waste water sludge. There is no other system available at present in the area.

REFERENCES

1. Instituto Nacional de Estadística, Geografía e Informática. (in Spanish). 2010. Available online: http//:www3.inegi.org.mx/sistemas/mexicocifras/default. aspx?e=31#P (accessed on 8 July 2013).
2. Duch–Gary, J. La Conformación Territorial del Estado de Yucatán; (in Spanish). Universidad Autónoma de Chapingo: Centro Regional de la Península de Yucatán, Mexico, 1988.
3. Magaña, A. Reactor Anaerobio Horizontal Doble (RAH-D) su desarrollo en la FI-UADY. (in Spanish). Ingeniería, Revista Académica de la Facultad de Ingeniería de la Universidad Autónoma de Yucatán. 2002, 6, 61–65.
4. Epstein, E. The Science of Composting; Technomic Publishing Company, Inc.: Lancaster, PA, USA, 1997.
5. Febles-Patrón, J.L.; Hoogesteijn, A. Evaluación preliminar de la eficiencia en las lagunas de oxidación de la ciudad de Mérida, Yucatán. (in Spanish). Ingeniería, Revista Académica de la Facultad de Ingeniería de la Universidad Autónoma de Yucatán 2010, 14, 127–137.
6. Haug, T.R. The Practical Handbook of Compost Engineering; Lewis Publishers: Boca Ratón, FL, USA, 1993.
7. Díaz, L.; Savage, G.M.; Eggerth, L.; Golueke, C. Composting and Recycling Municipal Solid Waste; Lewis Publishers: Boca Ratón, FL, USA, 1993.
8. Das, K.; Tollner, E.W.; Eiteman, M.A. Comparison of Synthetic and Natural Bulking Agents in Food Waste Composting. Compost Sci. Util. 2003, 11, 27–35.
9. Manios, V.I.; Stentiford, E.I.; Kefaki, M.D.; Siminis, C.I.; Dialynas, G.; Manios, T. Development of a Methodology for the Composting of Sludge in Crete; Grece. In Proceedings of the International Conference of Organic Recovery and Biological Treatment into the Next Millennium, Harrogate, UK, 3–5 September 1997.
10. Velazquez Aradillas, J.C.; Sauri Riancho, M.R.; Castillo Borges, E.R.; Herrera, J.; Alcocer-Vidal, V.; Estrella–May, F.; Cabañas–Vargas, D.D. Composting Of Septic Sludge. In Proceedings of the International Conference ORBIT 2006, Biological Waste Management: From Local to Global, Waimar, Germany, 13–15 September 2006.
11. Norma Oficial Mexicana NOM-004-SEMARNAT-2002, Protección ambiental. Lodos y biosolidos. Especificaciones y Límites máximos permisibles de contaminantes para su aprovechamiento y disposición final. (in Spanish). Available online: http://www.semarnat.gob.mx/leyesynormas/Pages/inicio.aspx (accessed on 8 July 2013).
12. Moodley, P.; Colleen, A.; Hawksworth, D. Standards Methods for the Recovery and Enumeration of Helminth ova in Wastewater, Sludge, Compost and Urine–Diversion Waste in South Africa; University of KwaZulu-Natal, Pollution Research Group Report No. TT322/08. WRC: Durban, South Africa, 2008.
13. Department of Civil Engineering; Water and Environmental Engineering Group, Standard Laboratory Procedures for the Analysis of Compost, University of Leeds: Leeds, UK, 2000.

Composting Used as a Low Cost Method for Pathogen Elimination 61

14. Zucconi, F.; Pera, A.; Forte, M.; de Bertoldi, M. Evaluating toxicity of inmature compost. BioCycle 1981, 22, 54–57.

15. V FCQAO, Method Book for the Analyses of Compost. Abfall Now e.V; Publishing House: Stuttgart, Germany, 1994; pp. 16–19.

16. Barrios, J.A.; Jiménez, B.; González, O.; Salgado, G.; Sanabria, L.; Iturbe, R. Destrucción de coliformes fecales y huevos de helmintos en lodos fisicoquímicos por vía ácida. (in Spanish). In Proceedings of the XII National Conference 2000: "Ciencia y Conciencia, Compromiso Nacional con el Medio Ambiente", Federación Mexicana deIngeniería Sanitaria y Ciencias Ambientales, Morelia, Mexico, 15–17 April 2000.

17. Pecson, B.M.; Nelson, K.L. Inactivation of Ascaris suum eggs by ammonia. Environ. Sci. Technol. 2005, 39, 7909–7914.

18. Baddi, G.A.; Albuquerque, J.A.; Gonzálvez, J.; Cegara, J.; Hafidi, M. Chemical and spectroscopic analysis of organic matter transformations during composting of olive mill wastes. Int. Biodeterior. Biodegrad. 2004, 54, 39–44.

19. Manser, A.; Keeling, A. Practical Handbook of Processing and Recycling Municipal Waste; CRC Press/Lewis Publishers: Boca Raton, FL, USA, 1996.

20. Cabañas-Vargas, D.D.; Sánchez-Monedero, M.A.; Urpilainen Kamilaki, A.; Stentiford, E.I. Assessing the stability and maturity of compost at large-scale plants. In Ingeniería, Revista Académica de la Facultad de Ingeniería de la Universidad Autónoma de Yucatán; 2005; Volume 9, pp. 25–30.

CHAPTER 4

An Experimental Investigation of Sewage Sludge Gasification in a Fluidized Bed Reactor

L. F. CALVO, A. I. GARCÍA, AND M. OTERO

4.1 INTRODUCTION

Sewage sludge originates from the process of treatment of municipal wastewater. In parallel with the enhancement and increase of sewage treatment plants to comply with more and more exigent environmental policies, production of sewage sludge has also increased and it is expected to increase even more. Although sludge treatment and disposal should be considered as an integral part of wastewater treatment, its handling is still one of the most significant challenges in wastewater management [1].

During the last years, former options for sewage sludge disposal, including ocean dumping, landfill, or disposal on agricultural land, have been forbidden, restricted, or became less acceptable from an environmental point of view. Consequently, cost-effective and environmentally friendly alternatives to these disposal means are needed. There is not

---

© *2013 by the authors.* The Scientific World Journal *Volume 2013 (2013), Article ID 479403; http:// dx.doi.org/10.1155/2013/479403. Creative Commons Attribution license (http://creativecommons. org/licenses/by/3.0/). Used with the authors' permission.*

an only and best management option for sewage sludge but it must be planned according to each sludge properties and local circumstances. Thus, under certain conditions, thermal processes may be appropriate options since they can be used for the conversion of large quantities of sewage sludge into useful energy [2]. Among thermal processes, gasification, that is, the thermal conversion of sewage sludge to combustible gas and ashes under a net reducing atmosphere, provides an attractive alternative to the most extended incineration [3]. Gasification not only accounts with all the advantages of incineration regarding sewage sludge management, such as complete sterilization, large volume reduction, or odour minimization, but also circumvents its main problems. The need for supplemental fuel, emissions of sulfur oxides, nitrogen oxides, heavy metals, and fly ash, and the potential production of dioxins and furans, which may be produced during incineration as a consequence of the oxidizing atmosphere [4], are avoided by sewage sludge gasification. A main advantage of sewage sludge gasification is that a high-quality flammable gas may be obtained, so it can be directly used for electricity generation or for supporting the drying of sewage sludge [5] and also may be employed as raw material in chemical synthesis processes [6]. Indeed, as for biomass gasification, a main difficulty about sewage sludge gasification is the presence of tar and dust in the synthesis gas produced, which may cause problems in process equipment and/or in turbines and engines for gas distribution [7–11]. In any case, gasification is considered a waste-to-clean energy technology [12].

Theory on sewage sludge gasification and a description of the process were reported by Dogru et al. [13] and also by Fytili and Zabaniotou [1] in their review on thermal processing of sewage sludge. Since Garcia-Bacaicoa et al. [14] first published a work on sewage sludge gasification, encouraging results have been reported, mainly during the last decade [3, 5, 6, 9–11, 15–18]. However, as recently highlighted by de Andrés et al. [6], comparatively with the large number of references that can be found in scientific literature regarding biomass gasification, experimental works on sewage sludge are relatively scarce. Thus, the overall objective of this work was to identify characteristics of sewage sludge gasification in atmospheric fluidized-bed gasifier. To this end, analyses of the mass transformation efficiency, gas composition, cold gas efficiency, and tar production were carried out.

4.2 MATERIALS AND METHODS

4.2.1. FUEL CHARACTERIZATION

Sewage sludge obtained from the Oakland (California) sewage treatment plant was dried, granulated, and stored in an airtight container until use. Sieve distribution of homogenized dried granulated sewage sludge is shown in Table 1. As can be observed, the most frequent size is 40 < mesh < 20, which accounts for 38.12%, particles larger than 40 mesh representing 22.4% of the sample.

Table 1. Sieve distribution and properties of sewage sludge.

	% Total
Particle size (mesh)	
>200	3.57
200 < 100	2.30
100 < 40	16.52
40 < 20	38.12
20 < 14	22.04
<14	17.64
Elemental analysis	
C (%)	36.2
H (%)	4.5
N (%)	5.6
S (%)	1.1
Cl (%)	0.1
O (%)	14.7
Proximate analysis	
Moisture (%)	7.9
Ash (%)	37.9
Volatiles (%)	55.1
Fixed carbon (%)	7.1
Heating value	
HHV[1] (MJ/kg)	15.4

[1] HHV: high heating value.
Except moisture, all values are on dry basis.

Before gasification, sewage sludge was analyzed for the following properties.

Moisture content was determined gravimetrically by the oven-drying method (ASTM D 3173, ASTM E 871). Triplicate samples, typically weighing 20 to 80 g each, were obtained from the container and air-dried at °C in an atmospheric oven to constant weight, normally obtained within 24 hours.

Higher heating value at constant volume (HHV) was measured using an adiabatic oxygen bomb calorimeter (via the equivalent methods ASTM D 2015, ASTM E 711, or ASTM D 5468). Triplicate samples of approximately 1 g each were split from batches prior to each test and analyzed using the Parr Model 1241 calorimeter with Model 1720 controller. Fuel was sampled in 1 g amounts, pelletized in a hand press to 12.7 mm diameter, and oven-dried to constant weight at °C prior to analysis.

Proximate determinations were made according to modified procedures from ASTM D 3172 through D 3175 (Standard Practice for Proximate Analysis of Coal and Coke); E 870 (Standard Methods for Analysis of Wood Fuels), D 1102 (ash in wood), and E 872 (volatile matter in wood); and the methods for refuse derived fuel (RDF)—E 830 (ash) and E 897 (volatile matter).

Triplicate samples, approximately 1 g each, split from the main sample batch were dried at °C and analyzed. Ash concentration was determined at 575°C for 2 hours in an atmospheric pressure air muffle. This temperature is that specified by ASTM for RDF and is slightly below the minimum temperature specified for wood (580°C).

Volatile concentration was determined under anaerobic conditions using a modified method for sparkling fuels in which samples in covered nichrome crucibles were placed in the front part of the open muffle furnace preheated to 950°C for 6 minutes to dispel volatiles over a period of more gradual heating and then brought to completion in the closed furnace during an additional 6 minutes, removed, and cooled under desiccant while still covered and weighed immediately.

Percent fixed carbon (dry basis) was computed by subtracting percent ash (dry basis) plus percent volatile matter (dry basis) from 100.

All crucibles were prefired at the test temperature (575 or 950°C) before use in order to remove any moisture or volatiles prior to each determination.

Experimental Investigation of Sewage Sludge Gasification   67

Ultimate analysis was carried out following standard methods.

4.2.2 GASIFICATION TESTS

4.2.2.1 GASIFICATION REACTOR

The reactor used for sludge gasification, which is schematically shown in Figure 1, is an atmospheric pressure rig, with a main reactor column of 73 mm inside diameter and 1 m in length. The main column discharges into a 127 mm^2 disengagement zone for internal recirculation of particles. Sewage sludge was fed to the reactor at a controlled rate using a custom-designed belt feeder driven by variable speed stepper motor. Fuel was injected in bed using a high-speed stainless steel auger. Fuel metering was controlled by the speed of the belt, while the auger was used solely for fuel injection. The fuel feeder and hopper were pressurized using a small amount of purge air (called secondary air) to prevent back-flow of reaction products into the fuel feeder.

The reactor column is made of 321 stainless steel and is surrounded by an electric furnace used to preheat the reactor. The electric furnace is automatically controlled using reactor wall temperature.

Fluidizing, or primary, air is preheated through a series of parallel electric heaters before being discharged through the distributor nozzles in the bottom of the bed. The bottom of the reactor terminates in a blind flange through which bed discharge, thermocouple, and pressure taps are inserted. The reactor was constructed in such a way that it could be rapidly disassembled for inspection and cleaning.

Above the furnace, the reactor expands into the disengagement section with four times the cross-sectional area of the main bed. Larger fuel and bed media particles are disengaged from the gas flow at this point and returned to the bed along the wall of the reactor column. Situated at the top of the disengagement section is a removable lid through which temperature, pressure, and bed make-up taps are inserted.

After the disengagement section the flow turns 90°. An ash drop-out is located at this position. A cyclone is situated past the horizontal pass and discharges separated particles through the bottom. Gas is flared at the

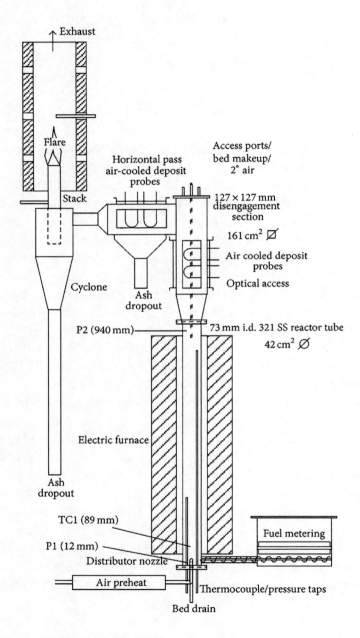

Figure 1. Fluidized-bed reactor (TC thermocouple, P pressure tap, and D disengagement thermocouple).

Experimental Investigation of Sewage Sludge Gasification

cyclone exit stack under natural aspiration inside a refractory lined exhaust duct ported to receive sample inlets. Gas and fly-ash samples are drawn from the cyclone exit.

Thermocouples, pressure transducers, outputs from continuous gas analyzers, and other electronic transducers are automatically recorded using a multichannel data logger communicating with a personal computer. A ten-second sampling interval is used.

4.2.2.2 BED MATERIAL

Bed material selected for the experiments was alumina-silicate sand (NARCO Investocast 60 grain). Fresh, screened bed material was used for each test. The required amount of fresh bed media was introduced in the gasifier before the gasification experiments. The initial mass of the bed was weighed and recorded. In any case, fresh media can be added through a top access port during operation and, also, bed material may be removed along gasification. After each test, spent bed was removed by dropping the lower flange plate and bed was captured. Any residue in the bed was determined from loss on ignition in an air muffle furnace at 575°C.

4.2.2.3 GAS ANALYSIS

Continuous gas sampling/analysis was accomplished by a Leeds & Northup gas analyzer (analyzes CO, CO_2, and H_2) and by a Panametrics O_2 analyzer for identifying the possible existence of leaks in the sample.

Grab samples were collected in glass for permanent gases, primarily CH_4 via GC. Primarily sampling locations for gases were taken at the cyclone exit.

4.2.2.4 ALKALI VAPORS SAMPLING

The alkali sampling train is schematically shown in Figure 2(a). The sample was extracted through a heated 5 m sintered stainless steel filter to separate particles. The filter was maintained at the same temperature as the gas at the sampling point so as to reduce errors in the determination of the alkali partitioning between the gas and particle phases. Filtered gas passed through a water-jacketed condenser and ice-bath cooled impinger train, a desiccant pump, a rotameter or mass flow meter, and a dry-test meter. This liquid volume was recorded and a sample was analyzed for the species of interest.

4.2.2.5 AMMONIA SAMPLING

Nitrogenous species other than NO_x (principally ammonia) were also measured via an absorption train. Ammonia sampling was conducted according to the methods of Ishimura [19], Furman et al. [20], and Blair et al. [21]. Similar to the alkali sampling, sample stream was drawn through a water-jacketed condenser and then through a set of ice-bath impingers filled with sulfuric acid 0.1 M solution. After the test, liquid volume from the impingers was weighed and recorded. An ion selective electrode (sensitive to ammonium) was used to measure (NH^{4+}) and hence ammonia. This train is schematically shown in Figure 2(b).

4.2.2.6 TAR

Tar was condensed in a set of condensers using dry-ice in ethanol to provide the low temperatures required. Methanol scrubbers were used after the condensers to capture some of the lighter hydrocarbons. The gas volume and tar weight were measured. Figure 2(c) schematically shows this train.

4.2.2.7 GENERALIZED TEST PROCEDURES

Sewage sludge was injected into the preheated bed at a rate controlled by the speed of the belt on the fuel feeder. Reactor heating was controlled automatically by the electric furnace around the reactor. Temperatures and pressures were monitored throughout each test. Total air and fuel flow

Experimental Investigation of Sewage Sludge Gasification

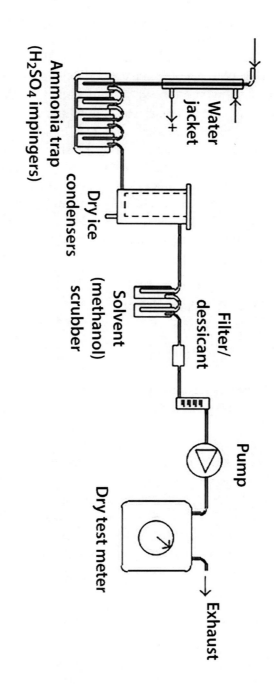

Figure 2. (a) Alkali sampling train; (b) ammonia sampling train; (c) tar sampling train.

rates were monitored, as were total flows through each of the sampling trains. Grab samples of gas were taken at frequent intervals for gas chromatography analysis. After gas sampling was finished, the fuel and air supplies were cut off from the reactor, and the reactor was cooled.

Posttest sampling was carried out after cooling of the reactor. The entire bed was recovered through the bottom reactor flange and analyzed for ashes and volatiles. Ash/char collected from the horizontal pass and cyclone was collected and put at the same test. Full mass and energy balances were completed as a check on analysis quality and to provide information about material and energy conversion efficiencies. Deposit probes were removed and depositions of the line samples were analyzed for the intended species after recording their weight. Ammonia and alkali were analyzed using an ionspecific electrode. Tar production was determined gravimetrically.

4.3 RESULTS AND DISCUSSION

4.3.1 FUEL PROPERTIES

Properties of the sludge used in this work are shown in Table 1. Properties of sewage sludge depend on sludge treatment and on sludge origin. Anyway, compared with sludge tested by other authors [6, 9–11, 15, 18], it can be observed that values in Table 1 are in the same order. Although ashes content is lower [6, 9–11] or similar [22] to that of some sewage sludge already used for gasification experimentation, it is high compared with that of biomass normally used for gasification. Volatiles content is in the range of the values found in the literature for sewage sludge to be used as fuel for gasification [10, 13, 22] while different biomass-based fuels used for gasification in the literature may have higher [23–25] or lower [26] volatiles content.

4.3.2 GASIFICATION TESTS

Since gasification is a partial combustion, the process requires less oxygen than a complete combustion. Considering the sludge chemistry analysis

Experimental Investigation of Sewage Sludge Gasification

shown in Table 1, a sludge elementary molecule $(C:H:O)$ could be written as $CH_{1.48}O_{0.31}$ and, then, complete combustion would be represented by the following equation:

$$CH_{1.48}O_{0.31} + 1.215(O_2 + 3.76N_2)$$
$$\rightarrow CO_2 + 0.74H_2O + 4.568N_2 \qquad (1)$$
$$\Delta H_f^0 = -422.03 kJ/mol$$

The air mass : fuel mass ratio (A/F) is 6.08 for complete combustion. For gasification, that ratio should be, in theory, between 0.2 and 0.4 [27]. Air flow and fuel feed rate were the parameters changed for experimentally obtaining a $0.2 < A/F < 0.4$. Two tests were carried out, test I and test II, Table 2 showing the corresponding main gasification parameters. A complete analysis for these tests was done, including gas analysis.

During test I, the accumulation of material inside the reactor was a main problem. As it may be seen in Table 2, the difference between bed media weight after and before test I was 1589.74 g. In order to avoid material accumulation and to improve the process heat transfer and, therefore, gasification efficiency, bed media was removed and fresh bed added along test II. As a result, there were no any accumulation problems during test II.

Ash and volatile analysis was carried out for depositions from the horizontal pass and cyclone and for the removed bed. Table 3 shows the obtained results. The ash content of those depositions was similar for tests I and II, the percentages being 85.05 and 80.91%, respectively. These values indicate that there were fuel particles in the removed bed, although less in test II than in test I because of having removed bed material during test II. Volatiles content is similar in the samples taken from cyclone and removed bed for both experiments. However, in the samples from horizontal pass, the percentage of volatiles in test I is 1.6 higher than in test II. This is due to the high content of not completely burned fuel particles inside the reactor in test I.

The mass transformation efficiency was determined by a mass balance and results are shown in Table 4. Percentage of raw produced gas is higher in test II than in test I, with values of 79.69 and 40.30%, respectively. That means that removed bed material during the test is convenient for this fuel if retention of particles inside the reactor want to be avoided. Depositions from pass horizontal are higher in test I for the same reason.

Table 2. Gasification parameters in tests I and II.

	Test I	Test II
Wet fuel burned (g)	4521.7	11743.9
Wet fuel feed rate (g/s)	8.94	12.52
Reactor preheating (°C)	850	850
Primary air preheating (°C)	300	300
Primary air (L/min)	15	20
Bed material	Al-Si	Al-Si
Fresh bed material (g)	866	433
Removed bed material (g)	2455.74	$433 + 128.6 + 1782.6^1$
Ash horizontal pass (g)	1261.6	193.1
Ash cyclone (g)	115.3	113.2
Alkali sampling train		
Bottle 1 (g)	46.6	97.1
Bottle 2 (g)	3.8	50.7
Bottle 3 (g)	0.9	−121.5
Bottle 4 (g)	0.5	0.5
Desiccant (g)	1.9	2.8
Filter paper (g)	0.0947	0.052
Filter cake (g)	3.3	0.4
Gas analyzed (m^3)	0.0972	0.1151
Time in alkali train (min)	26	123
Average flow rate through tar train ($L\,min^{-1}$)	3.738	0.936
Ammonia sampling train		
Bottle 1 (g)	109.9	85.6
Bottle 2 (g)	11.5	8.3
Bottle 3 (g)	2.3	2.1
Bottle 4 (g)	0.7	0.1
Desiccant (g)	2.9	6
Filter paper (g)	0.0066	0.0003
Gas analyzed (ft^3)	4.631	14.101
Time in ammonia sampling train (min)	26	123
Average flow rate through ammonia sampling train ($L\,min^{-1}$)	5.044	3.246

Experimental Investigation of Sewage Sludge Gasification

Table 2. Continued.

Tar sampling train		
Condenser 1 (g)	180.3	147.3
Bottle and tubing (g)	94.6	92.4
Line 1 (g)	2	3.5
Condenser 2 (g)	75.2	113.6
Line 2 (g)	0.7	1.4
Impinger 1 (g)	−6.8	−219
Impinger 2 (g)	7.8	3.2
Line 3 (g)	0	0.1
Desiccator (g)	2.6	12.5
Filter paper (g)	0.0197	0.0025
Gas analyzed (m^3)	0.3955	0.7547
Time in tar train (min) Average flow rate through	44	123
tar train (L min^{-1})	8.989	6.136

[1] Added bed during the test.

Produced gas was analyzed in tests I and II and Table 5 shows the average composition obtained.

The high value of nitrogen concentration was due to gasification using air. This means that, although the process is cheaper than with pure oxygen, the quality of the produced gas is lower. The sum of the H2, CO, and CH_4 percentages in tests I and II is 44 and 40.7%, respectively; these values are higher than values found in literature [16]. Seeing that the high heating values of CO, H_2, and CH_4 are 12.87, 12.99, and 41.22 MJ/Nm^3, respectively, it can be deduced that produced gas from test I was better, in terms of energetic quality, than that produced in test II, which can be owing to the fact that flow air was higher in test II.

For determining the energy efficiency of the system, an energy balance was calculated, results being shown in Table 4 together with the corresponding energetic efficiency parameters.

The heating value of the produced gas was 9.33 and 8.42 MJ/m^3, in tests I and II, respectively, values which are quite close. However, it must be taken into account that, although heating value of the produced gas is

Table 3. Ash and volatile analysis of depositions from horizontal pass, from cyclone, and from removed bed for tests I and II.

	Ash from horizontal pass deposition (%)	Ash from cyclone depositions (%)	Ash from removed bed (%)	Volatiles from horizontal pass deposition (%)	Volatiles from cyclone depositions (%)	Volatiles from removed bed (%)
Test I	80.84	79.66	85.05	9.10	13.80	4.08
Test II	82.43	79.15	80.91	5.64	12.49	4.94

slightly higher in test I, the mass gas obtained in this test is almost twice lower, with respect to inputs, than in test II. Common values for cold gas efficiency are between 0.45 and 0.67 [28]. This range was attained in test II, but not in test I, which may be related to the accumulation of particles inside the reactor and derived problems.

For test I, 87.6% of the gas energy was obtained due to its chemical energy and the most important contribution was from methane (42.52%). Although hydrogen is the component that presents the highest enthalpy (141.90 MJ/kg), its contribution in chemical energy was only of 34.07% because, as may be seen in Table 5, its volume percentage was 21%. For test II, chemical energy is equal to 81.4% of total energy. In this case the most important contribution was also from methane, with 38.16%. Finally, the lost energies in test I were higher (61% with respect to inputs) due to the same reasons explained before.

4.3.3 TAR TRAIN

Dust, water, and tars constitute depositions of tar train. Dust percentage in produced gas is known by the depositions in the heater filter placed in alkali train. The difference in weight before and after the train is the dust accumulated on it.

The tar quantities condensed in the tar train during tests I and II are shown in Table 5. The mass of tar from test I was 1.4 times higher than that from test II. The total tar values, with respect to the sewage sludge fed, are

Experimental Investigation of Sewage Sludge Gasification

Table 4. Mass balance, energy balance, and energetic efficiency parameters for tests I and II.

	Test I		Test II	
	Mass (g)	Input %	Mass (g)	Input %
Mass balance				
Inputs				
m_a	1032.54	16.08	3618.27	22.91
m_{df}	4131.97	64.36	10733.92	67.96
m_m	389.73	6.07	1009.98	6.39
m_h	866	13.49	433	2.74
Outputs				
m_{hp}	126.16	19.65	238.06	1.51
m_c	115.3	1.80	114.57	0.73
m_{rc}	2455.79	38.25	2854	18.07
m_g	2587.55	40.3	12588.54	79.69
	Test I		Test II	
	Energy (MJ)	Power (W)	Energy (MJ)	Power (W)
Energy balance				
Inputs				
Fuel	64.0282	24253	166.2960	22533.3
Air	0.1745	66.1	0.6429	87.1
Heaters	1.2804	485	1.6421	222.5
Outputs				
Lost	39.2756	14877.1	51.8074	7020.0
Gas	26.3075	9927.1	116.581	15822.9
	Test I		Test II	
Energetic efficiency				
Hot gas efficiency (HGE)	0.41		0.70	
Cold gas efficiency (CGE)	0.34		0.57	
% lost with respect to inputs	0.60		0.31	
% lost with respect to fuel	0.61		0.31	

m_a: air mass; m_{df} dry fuel mass; m_m: fuel moisture mass; m_h: fresh bed mass; m_{hp}: ash from horizontal pass mass; m_c: ash from cyclone mass; m_{rc}: removed bed mass; m_g: raw produced gas mass.

Table 5. Average concentration (%v/v) of O_2, CO, CO_2, H_2, CH_4 and N_2 in produced gas from tests I and II and tar produced in each test.

	Test I	Test II
Produced gas		
O_2	0.6	0.4
CO	14.6	13.4
CO_2	11.4	11.1
H_2	21.0	20.7
CH_4	8.4	6.7
N_2	34.1	36.0
Total	90.1	88.2
Produced tar		
Total depositions (g)	283.75	289.07
Dust (g)	12.21	2.62
% moisture	95.58	98.94
Total tar (g)	0.3348	0.4413
Tars (g/m³)	0.846	0.585

comparatively lower than those obtained by [22] during the gasification of sewage sludge in a spouted bed reactor. In any case, a gas cleaning system for tar removal would be needed for industrial applications. Between wet and hot gas cleaning, the latter should be preferred since it really destroys the tars instead of transferring them to a liquid phase, which would need further and expensive treatment [29]. Also, for energy efficiency reasons of the whole process, the hot gas cleaning is a promising method [30] and several authors have recently studied this sort of method with successful results [30–33].

4.4 CONCLUSIONS

Gasification of sewage sludge was carried out using air, which is cheaper than pure oxygen. A high quality gas (H_2, CO, and CH_4 summed up to

Experimental Investigation of Sewage Sludge Gasification

40.7–44%) with a heating value of 8.42–9.33 MJ/Nm^3, low in tar content (0.6 g/m^3) and cold gas efficiency of 57% was obtained. Comparatively with published results on the gasification of sewage sludge [34], these values are outstanding and demonstrate that gasification of sewage sludge, which may be carried out in a real straightforward way, is an option for the valorization of sewage sludge.

REFERENCES

1. D. Fytili and A. Zabaniotou, "Utilization of sewage sludge in EU application of old and new methods—a review," Renewable and Sustainable Energy Reviews, vol. 12, no. 1, pp. 116–140, 2008.
2. S. Werle and R. K. Wilk, "A review of methods for the thermal utilization of sewage sludge: the Polish perspective," Renewable Energy, vol. 35, no. 9, pp. 1914–1919, 2010.
3. B. McAuley, J. Kunkal, and S. E. Manahan, "A new process for the drying and gasification of sewage sludge," Water Engineering and Management, vol. 148, no. 5, pp. 18–22, 2001.
4. M. Jaeger and M. Mayer, "The Noell conversion process—a gasification process for the pollutant-free disposal of sewage sludge and the recovery of energy and materials," Water Science and Technology, vol. 41, no. 8, pp. 37–44, 2000.
5. C. J. Hamilton, "Gasification as an innovative method of sewage sludge disposal," Journal of the Chartered Institution of Water and Environmental Management, vol. 14, no. 2, pp. 89–93, 2000.
6. J. M. de Andrés, A. Narros, and M. E. Rodríguez, "Behaviour of dolomite, olivine and alumina as primary catalysts in air-steam gasification of sewage sludge," Fuel, vol. 90, no. 2, pp. 521–527, 2011.
7. P. Simell, E. Kurkela, and P. Stahlberg, "Formation and catalytic decomposition of tars from fluidized-bed gasification," in Advances in Thermochemical Biomass Conversion, M. Bridgwater, Ed., pp. 265–279, 1993.
8. L. Devi, K. J. Ptasinski, and F. J. J. G. Janssen, "A review of the primary measures for tar elimination in biomass gasification processes," Biomass and Bioenergy, vol. 24, no. 2, pp. 125–140, 2002.
9. I. Petersen and J. Werther, "Experimental investigation and modeling of gasification of sewage sludge in the circulating fluidized bed," Chemical Engineering and Processing: Process Intensification, vol. 44, no. 7, pp. 717–736, 2005.
10. J. J. Manyà, J. L. Sánchez, J. Ábrego, A. Gonzalo, and J. Arauzo, "Influence of gas residence time and air ratio on the air gasification of dried sewage sludge in a bubbling fluidised bed," Fuel, vol. 85, no. 14-15, pp. 2027–2033, 2006.
11. J. M. de Andrés, A. Narros, and M. E. Rodríguez, "Air-steam gasification of sewage sludge in a bubbling bed reactor: effect of alumina as a primary catalyst," Fuel Processing Technology, vol. 92, no. 3, pp. 433–440, 2011.

12. N. Nipattummakul, I. I. Ahmed, S. Kerdsuwan, and A. K. Gupta, "Hydrogen and syngas production from sewage sludge via steam gasification," International Journal of Hydrogen Energy, vol. 35, no. 21, pp. 11738–11745, 2010.

13. M. Dogru, A. Midilli, and C. R. Howarth, "Gasification of sewage sludge using a throated downdraft gasifier and uncertainty analysis," Fuel Processing Technology, vol. 75, no. 1, pp. 55–82, 2002.

14. P. Garcia-Bacaicoa, R. Bilbao, and C. Uson, "Sewage sludge gasification: first studies," in Proceedings of the 2nd Biomass Conference: Energy, Environment, and Agricultural Industry, pp. 685–694, 1995.

15. A. van der Drift, J. van Doorn, and J. W. Vermeulen, "Ten residual biomass fuels for circulating fluidized-bed gasification," Biomass and Bioenergy, vol. 20, no. 1, pp. 45–56, 2001.

16. A. Midilli, M. Dogru, C. R. Howarth, M. J. Ling, and T. Ayhan, "Combustible gas production from sewage sludge with a downdraft gasifier," Energy Conversion and Management, vol. 42, no. 2, pp. 157–172, 2001.

17. F. Pinto, H. Lopes, R. N. André, M. Dias, I. Gulyurtlu, and I. Cabrita, "Effect of experimental conditions on gas quality and solids produced by sewage sludge cogasification. 1. Sewage sludge mixed with coal," Energy and Fuels, vol. 21, no. 5, pp. 2737–2745, 2007.

18. B. Groß, C. Eder, P. Grziwa, J. Horst, and K. Kimmerle, "Energy recovery from sewage sludge by means of fluidised bed gasification," Waste Management, vol. 28, no. 10, pp. 1819–1826, 2008.

19. D. M. Ishimura, Investigation of nitrogenous compound formation in biomass gasification [MSME thesis], University of Hawaii, 1994.

20. A. H. Furman, S. G. Kimura, R. E. Ayah, and J. F. Joyce, "Biomass gasification pilot plant study," Final Report, Vermont DPS Contract No 0938222, General Electric Corporate Research and Development, Schenectady, NY, USA, 1992.

21. D. W. Blair, J. O. L. Wendt, and W. Bartok, "Evolution of nitrogen and other species during controlled pyrolysis of coal," Symposium (International) on Combustion, vol. 16, no. 1, pp. 475–489, 1977.

22. A. Adegoroye, N. Paterson, X. Li et al., "The characterisation of tars produced during the gasification of sewage sludge in a spouted bed reactor," Fuel, vol. 83, no. 14-15, pp. 1949–1960, 2004.

23. L. Shen, Y. Gao, and J. Xiao, "Simulation of hydrogen production from biomass gasification in interconnected fluidized beds," Biomass and Bioenergy, vol. 32, no. 2, pp. 120–127, 2008.

24. D. Vera, F. Jurado, and J. Carpio, "Study of a downdraft gasifier and externally fired gas turbine for olive industry wastes," Fuel Processing Technology, vol. 92, no. 10, pp. 1970–1979, 2011.

25. J. D. Martínez, E. E. S. Lora, R. V. Andrade, and R. L. Jaén, "Experimental study on biomass gasification in a double air stage downdraft reactor," Biomass and Bioenergy, vol. 35, no. 8, pp. 3465–3480, 2011.

26. N. Ramzan, A. Ashraf, S. Naveed, and A. Malik, "Simulation of hybrid biomass gasification using Aspen plus: a comparative performance analysis for food, municipal solid and poultry waste," Biomass and Bioenergy, vol. 35, no. 9, pp. 3962–3969, 2011.

Experimental Investigation of Sewage Sludge Gasification

27. D. Pfaff, B. Jenkins, and S. Turn, "Elemental gas-particle partitioning in fluidized bed combustion and gasification of a biomass fuel," in Progress in Thermochemical Biomass Conversion, A. V. Bridgwater, Ed., Blackwell Science, Oxford, UK, 2008.

28. P. Panaka, "Operating experience with biomass gasifiers in Indonesia," A. V. Bridgwater, Ed., pp. 392–402, Blackie, Glasgow, UK, 2004.

29. J. Corella, M. A. Caballero, M. P. Aznar, and J. Gil, "Biomass gasification with air in fluidized bed: hot gas cleanup and upgrading with steam reforming catalysts of big size," in Biomass: A Growth Opportunity in Green Energy and Value Added Products, R. P. Overend and E. Chornet, Eds., vol. 2, pp. 933–938, Elsevier Science, 1999.

30. S. Schmidt, S. Giesa, A. Drochner, and H. Vogel, "Catalytic tar removal from bio syngas—catalyst development and kinetic studies," Catalysis Today, vol. 175, no. 1, pp. 442–449, 2011.

31. H. Rönkkönen, P. Simell, M. Niemelä, and O. Krause, "Precious metal catalysts in the clean-up of biomass gasification gas—part 2: performance and sulfur tolerance of rhodium based catalysts," Fuel Processing Technology, vol. 92, no. 10, pp. 1881–1889, 2011.

32. M. Stemmler and M. Müller, "Chemical hot gas cleaning concept for the "CHRISGAS" process," Biomass and Bioenergy, vol. 35, no. 1, pp. S105–S115, 2011.

33. M. Stemmler, A. Tamburro, and M. Müller, "Laboratory investigations on chemical hot gas cleaning of inorganic trace elements for the "UNIQUE" process," Fuel, vol. 108, pp. 31–36, 2013.

34. P. Manara and A. Zabaniotou, "Towards sewage sludge based biofuels via thermochemical conversion—a review," Renewable and Sustainable Energy Reviews, vol. 16, no. 5, pp. 2566–2582, 2012.

CHAPTER 5

Bacterial Consortium and Axenic Cultures Isolated from Activated Sewage Sludge for Biodegradation of Imidazolium-Based Ionic Liquid

M. MARKIEWICZ, J. HENKE, A. BRILLOWSKA-DĄBROWSKA, S. STOLTE, J. ŁUCZAK, AND C. JUNGNICKEL

5.1 INTRODUCTION

Ionic liquids (ILs) are chemicals usually composed of large asymmetric, organic cation and organic or inorganic anion. Physical and chemical properties of this group of compounds can vary significantly what allows them to be designed for a particular purpose (Krossing et al. 2006). The last decade has shown a growing interest in the application of ILs in gas storage and separation, catalysis, electrodeposition of metals, waste and biomass reprocessing, energy production, etc. (Kragl et al. 2002; Jiang et al. 2006; Plechkova and Seddon 2008). When applied in such industrial processes, ILs will inevitably emerge in wastewaters and might end up in natural soils or water bodies by breaking through treatment systems or due to the accidental release during transport and storage. Although, the low volatility of ILs can be an advantage in reducing air emissions and thereby decreasing the risk of human exposure, the relatively high toxicity

© 2013 by the authors; licensee Springer. International Journal of Environmental Science and Technology (2013) 11:390, DOI: 10.1007/s13762-013-0390-1. Creative Commons Attribution license (http://creativecommons.org/licenses/by/3.0/).

and resistance to biotic and abiotic degradation that could be observed for some of the ILs structures is a concern (Romero et al. 2008).

Biodegradation of substituted imidazolium cation was examined in detail by a number of research groups (Gathergood et al. 2006; Romero et al. 2008; Stolte et al. 2008; Markiewicz et al. 2009; Abrusci et al. 2010; Coleman and Gathergood 2010). Stolte et al. conducted a comprehensive study of biodegradation of 1-methyl-3-alkylimidazolium chlorides. No primary biodegradation was observed for methyl- to butyl-substituted compounds, even those containing oxygen atoms introduced into the alkyl chain (as ethers or terminal hydroxyl groups) which are known to increase biodegradability. Better biodegradability was observed for higher homologs with 1-methyl-3-octylimidazolium chloride reaching 100 % biodegradation after 24 day of the test (Stolte et al. 2008). In most IL biodegradation tests, activated sewage sludge is used as an inoculum as it is composed of multiple species with a wide taxonomic diversity and is therefore more likely to contain specie or species capable of degrading ILs (Gathergood et al. 2006; Romero et al. 2008; Stolte et al. 2008; Markiewicz et al. 2009). In the previously mentioned work, Stolte et al. examined the ability of commercially available freeze-dried bacteria mixture to biodegrade ILs and concluded that it was unable to degrade any of tested compounds. Modelli et al. used inoculum derived from top soil for biodegradation of ILs under ASTM D 5988-96 test conditions and found it capable of degrading 1-butyl-3-methylimidazolium chloride (Modelli et al. 2008). Abrusci et al. used a strain of *Sphingomonas paucimobilis* isolated from cinematographic film for biodegradation of some common ILs and found it predisposed to degrade many tested compounds including those previously reported to be non-biodegradable. However, a somehow surprising trend was observed by Abrusci et al. for imidazolium chlorides showing highest biodegradation for ethyl- and lowest for octyl- and decyl-substituted compounds (Abrusci et al. 2010).

Some bacterial species are capable of tolerating or even growing in the presence of xenobiotics (Isken and deBont 1998; Takenaka et al. 2007). In pursuit of understanding the mechanisms of bacterial resistance and adaptation to xenobiotics, it was revealed that resistance is either a natural property of a species or is acquired by genetic changes (Weber and deBont 1996; Pham et al. 2009). The ability to degrade xenobiotics is not simply a function of the amount of biomass of inoculum, although very often a correlation between the two exists. Activated sewage sludge community is often used in biodegradation tests as it is expected that degraders of xenobiotics will be encountered among the multitude of species. Assuring this by adding specialized species is known as (bio)augmentation and was previously suggested as a promising strategy to enhance the degradative capacity of soils and sewage sludge (Limbergen et al. 1998; Pieper and Reineke 2000). For this reason, an attempt to identify the exact microbial strains which might be partaking in the process of biological degradation of ILs was undertaken here in order to uncover specie or species especially predisposed to degrade ILs. Would the attempt be successful, it might present a very useful tool in enhancing biodegradation of ILs by augmenting indigenous microbial communities. As activated sewage sludge was clearly proven to be the most potent in biodegrading ILs, it was chosen as a starting point. For the current test, 1-methyl-3-octylimidazolium chloride was selected since it was previously shown to be biodegradable in activated sewage sludge, and therefore, it will allow for comparative statements to be drawn. A nearly 30-fold increase in maximum biodegradable concentration and growing rates of degradation resulting from pre-exposition of activated sewage sludge community to this IL were reported before. Moreover, a break-down of the imidazolium ring was observed (Markiewicz et al. 2011). In the course of the current work, nine strains of bacteria from adapted activated sewage sludge were isolated, and the influence of 1-methyl-3-octylimidazolium chloride on their growth was tested. The performance of consortium of all isolated strains and of each strain individually was verified in 'Manometric respirometry' biodegradation test using the same IL. Results of these tests are presented hereby.

5.2 MATERIALS AND METHODS

5.2.1 CHEMICALS

The investigated ionic liquid 1-methyl-3-octylimidazolium chloride was purchased from Merck KGaA (Germany) with a purity of ≥ 98 %. The molecular weight, chemical formula and molecular structure are shown in Table 1.

5.2.2 SELECTION OF RESISTANT ISOLATES

The experiments were carried out at the Department of Chemical Technology, Gdańsk University of Technology and at the Center for Environmental Research and Sustainable Technology, University of Bremen. The sewage was sampled in June 2009, and the measurements were carried out in the subsequent 10 months. Activated sewage sludge was obtained from aeration tank of municipal wastewater treatment plant (WWTP) 'Wschód' in Gdańsk, Poland, and used for biodegradation experiment during which it was subjected to selective pressure of [OMIM][Cl] as described in Markiewicz et al. (2009). Subsequently, microbial strains were isolated and identified as described below.

5.2.3 BACTERIAL SPECIES IDENTIFICATION

Activated sewage sludge, obtained from the last phase of biodegradation test, was diluted with saline (0.85 % NaCl) solution. Samples diluted from 1 to 5 times were then inoculated into Petri dishes containing Luria Agar (LA) using spread-plate method and incubated in 16 °C for 48 h. To obtain pure cultures, single colonies were chosen and inoculated to Petri dishes with LA agar, using streak plate method and incubated. Single colonies were inoculated to test tubes with 2 mL of Luria Broth (LB) and cultivated. After 48 h, 100 µL of each liquid culture was inoculated into Petri dishes containing LA agar by spread-plate method and cultured for 48 h at 16 °C. Bacterial isolates from activated sewage sludge were subjected to Gram staining and observed under optical microscope.

Bacterial Consortium and Axenic Cultures Isolated from Activated Sewage Sludge   87

Table 1. Results including pathogens reduction in time and control parameter behaviour.

Name	Chemical formula	Molecular structure	Molecular mass (g/mol)
1-Methyl-3-octylimidazolium chloride	$C_{12}H_{23}ClN_2$		230.78

5.2.4 MOLECULAR IDENTIFICATION OF BACTERIAL ISOLATES

DNA was purified by mean of Genomic Mini Kit (A&A Biotechnology, Poland). PCR with universal 16S rDNA primers: 16S-for 5'GGACTAC-CAGGGTATCTAATC 3' and 16S-rev 5' GATCCTGGCTCAGGATGAAC 3' and REDTaq® ReadyMix PCR Reaction Mix (Sigma-Aldrich®) were performed with isolated DNA of nine bacterial strains in a volume of 20 µL. The time–temperature profile for PCR was 35 cycles of 30 s at 94 °C, 30 s at 55 °C and 45 s at 72 °C, preceded by initial denaturation for 10 min at 95 °C. The presence of specific PCR products of approximately 800 bp was examined using electrophoresis on a 1 % agarose gel and staining with ethidium bromide.

The PCR products were purified with High Pure PCR Product Purification Kit (Roche) and sequenced (Genomed, Poland). The identification of isolates was performed by comparison of obtained sequences against the sequences from GenBank (http://blast.ncbi.nlm.nih.gov)

5.2.5 GROWTH INHIBITION TEST

Fresh bacterial cultures of the nine isolates were grown in LB broth (cell density $4 \cdot 10^8$ cells/L). IL ([OMIM][Cl]) was used in four concentrations: 0.2; 2; 20 and 200 mM. Positive control samples, containing LB, inoculum and sodium glutamate with the same carbon content as respective samples with IL—one for each microorganism—were prepared. Additionally, the sample with LB and IL only served as a negative control. All prepared samples were then incubated in 16 °C for 48 h. After incubation, in order to determine the growth rate in each sample, 200 µL of each solution was transferred to the disposable microplate, and the quantity of bacterial cells was determined spectrophotometrically using a multilabel plate reader Wallac 1420 VICTOR[3]-V (λ = 595 nm wavelength).

5.2.6 BIODEGRADATION TESTS

Ready biodegradability tests of 1-methyl-3-octylimidazolium chloride in concentration of 0.2 mM using 1 mL of axenic cultures (resulting in cell density of $8 \cdot 10^4$ cells/L in each culture) of nine isolates were performed according to OECD 301 F 'Manometric respirometry' procedure (OECD 1992). In this test, biodegradation is measured as a decrease in pressure in gas-tight test vessel caused by depletion of oxygen used for aerobic degradation of IL reduced by the blank sample value (sample showing only endogenous respiration of bacteria, without addition of test compound) with respect to the theoretical amount of oxygen necessary to completely oxidize the compound tested. Since almost no biodegradation was observed, the same test was repeated for a consortium composed of all nine isolates (cell density $8 \cdot 10^5$ cells/L of all strains in total) in a concentration of IL previously reported to be low enough for biodegradation with activated sewage sludge to occur (0.25 mM) (Stolte et al. 2008). Samples containing activated sewage sludge (cell density $10 \cdot 10^7$ cells/L) derived from 'Wschód' in Gdańsk, were also employed in the test for the sake of comparison.

5.3 RESULTS AND DISCUSSION

5.3.1 ISOLATION AND IDENTIFICATION OF SEWAGE SLUDGE BACTERIA

Nine different isolates of microorganisms were cultivated from original biodegradation test utilizing activated sewage sludge. After DNA isolation, followed by 16S rDNA PCR and product purification, the DNA sequencing was conducted. The obtained sequences were compared with GenBank data (NCBI 2011), and microbial species were identified.

Table 2 summarizes the results of identification and Gram staining. Figure 1 presents the phylogenetic tree of identified species. The sequence alignment and bootstrap values were calculated using CLUSTALW2.012.

Table 2. Microbial isolates identified by sequencing of 16S rDNA sequences comparison with GenBank data (NCBI 2011).

No.	Organism	Gram
1	Flavobacterium sp. WB3.2-27	–
2	Shewanella putrefaciens CN-32	–
3	Moraxellaceae Bacterium MAG	–
4	Flavobacterium sp. FB7	–
5	Microbacterium keratanolyticum AO17b	+
6	Flavobacterium sp. WB 4.4-116	–
7	Arthrobacter sp. SPC 26	+
8	Rhodococcus sp. PN8	+
9	Arthrobacter protophormiae strain DSM 20168	+

5.3.2 GROWTH INHIBITION

The inhibition of growth of nine isolated bacterial strains by 1-methyl-3-octylimidazolium chloride [OMIM][Cl] in four concentrations covering four orders of magnitude was examined. Therefore, the strains were incubated in the presence of IL for 48 h, and afterward, the cell density was determined spectrophotometrically. The results were expressed as a percent of growth in relation to a positive control with sodium glutamate (Fig. 2).

Concentration-dependent decrease in growth was found. In 0.2 mM [OMIM][Cl], most of strains showed a similar decrease in growth (10–20 %) except strains 4 and 5 where growth reached approximately 150 and 120 % of positive control, respectively. This might suggest that those two species are especially predisposed to utilizing [OMIM][Cl] as carbon source or, more probably, might have been a result of hormesis. It is believed that when exposed to low doses of toxin, organisms exhibit increased growth by investing larger amounts of energy in reproduction in order to assure survival (Calabrese 2005). It is possible that for Flavobacterium sp. FB5

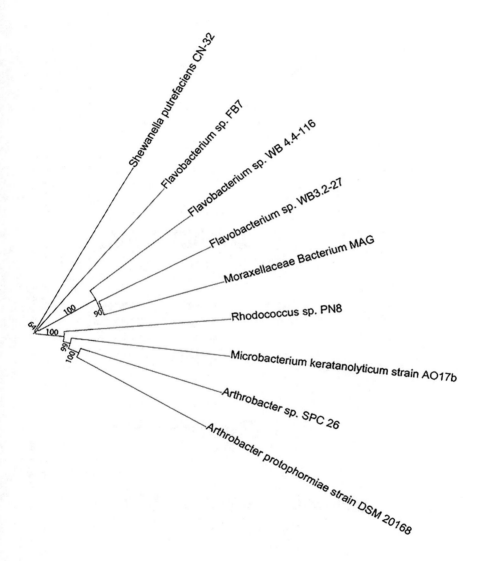

Figure 1. Phylogenetic tree of identified species (visualized using iTOL Letunic and Bork 2007).

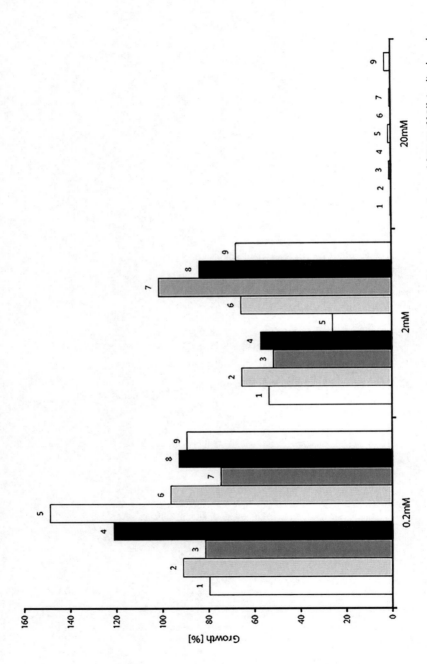

Figure 2. Growth of the selected bacterial strains in solution containing various concentrations of [OMIM][Cl] is displayed as a percent of positive control. Strains were numbered in accordance to Table 2.

and *Microbacterium keratanolyticum*, this concentration of IL triggered the hormetic effect.

Increasing IL concentration to 2 mM caused 50–60 % inhibition of growth for most of strains. *Microbacterium keratanolyticum* proved to be particularly sensitive (75 % inhibition) supporting hypothesis of hormesis. No inhibition of growth of *Arthrobacter* sp. SPC 26 and only minor inhibition of *Rhodococcus* sp. PN8 were found. Enhanced growth of *Arthrobacter* sp. as compared 2 mM sample can again be explained by hormesis. *Rhodococcus* exhibited steady growth in both 0.2 and 2 mM samples, making this strain the most tolerant for IL.

Most of isolated strains proved to be slightly more resistant to IL than previously reported by Łuczak et al. (2010), where at least 80 % growth inhibition was observed in 4 mM [OMIM][Cl] for Gram-negative Escherichia coli and in 2 mM [OMIM][Cl] for three other Gram-positive bacterial strains. The growth observed for samples containing 20 mM IL solution was almost negligible and completely inhibited in 200 mM solutions (results not shown).

A variety of negative effects of chemicals on function of bacterial cells are known. For example, membrane permeabilization is caused by adsorption of molecules, resulting in leakage of cytoplasm and micromolecules or simply inability to maintain cell integrity. Other effects include the decrease in the cell energy status caused by passive flux of ions through the membrane, the disturbance of functions of other (not involved in energy transduction) proteins present in the membrane as well as the distortions in fluidity and the hydration of the membrane surface (Isken and deBont 1998). Bacterial cells might counteract these effects by adaptation at the level of cytoplasmic membrane, including changes in level of saturation of fatty acids which changes the fluidity of membrane in order to compensate the negative effect of chemicals (known as homeoviscosic adaptation Heipieper and deBont 1994); reduction in cell hydrophobicity by altering the content of lipopolysaccharide (LPS) and modification of porines which was proven to increase the solvent tolerance of microorganism; solvents degradation into less- or nontoxic substances and the solvent active excretion from the cell (Isken and deBont 1998). It would be expected that bacteria able to survive in unfavorable conditions created by the presence of ionic liquids would employ one of these mechanisms in order to survive, yet not necessarily will be able to break them down. The difference in cell wall construction of Gram-positive and Gram-negative bacteria might be responsible for their slightly different resistance toward chemicals. Even though Gram-negative cell wall is much thinner, it is covered with an additional lipid membrane acting as a barrier for many biocides, and thus, Gram-negative microorganisms show lower sensitivity to organic chemicals including surfactants (Blazevic 1976; Volkering et al. 1995). The same regularity, though only slightly pronounced, was observed by Łuczak et al. for ILs (Luczak et al. 2010). Nevertheless, no obvious differences in growth inhibition between Gram-positive and Gram-negative bacteria were observed within this work.

5.3.3 BIODEGRADATION TESTS

All isolated stains from previous experiment (Markiewicz et al. 2009) were tested in biodegradation tests according to OECD 301 F (Manometric respirometry) in order to determine the mineralization of [OMIM][Cl]. None of the isolates exhibited any significant levels of biodegradation of 1-methyl-3-octylimidazolium chloride when applied as an axenic culture. A maximum value of 8 % was observed for *Microbacterium keratanolyticum* (results not shown) which was expected as this strain showed the highest growth rate in growth inhibition test in 0.2 mM sample. It is probable that not an axenic culture but a consortium of two or more strains is necessary to conduct full biodegradation of IL. Numerous examples of such situations can be found in literature. Consortia of bacteria were found to degrade azo-dyes, crude oil hydrocarbons, atrazine, etc., more efficiently than individual strains (Zwieten and Kennedy 1995; Khehra et al. 2005; Adebusoye et al. 2007). There are a number of interactions within a consortium that lead to degradation of a xenobiotic and that cannot occur in axenic culture. These are unfortunately very difficult to uncover. It may be that the first step is conducted by one organism resulting in an intermediate that is then transformed further by another organism, and this cascade proceeds to full mineralization. At some point, more than one organism can be involved in the degradation of intermediates leading to different products and consequent ramification of the metabolic pathway. It

Bacterial Consortium and Axenic Cultures Isolated from Activated Sewage Sludge   95

is also possible that some members of the consortia do not take part in the degradation of the xenobiotic, but support the primary degraders providing them with essential nutrients, e.g., vitamins, amino acids or creating appropriate environmental conditions such as removing toxins and adjusting oxygen levels. It is therefore far more likely that enhanced degradation can be achieved by augmentation with a consortium rather that an axenic culture (Grady 1985; Schink 2002).

The results of the biodegradation test with the consortium formed by mixing all nine isolates are shown in Fig. 3a. Biodegradation of sodium glutamate was also conducted to examine the viability of the inoculum. Additionally, biodegradation of both [OMIM][Cl] and sodium glutamate conducted by activated sewage sludge is displayed for comparison

It is clear that both microbial communities were viable as sodium glutamate was completely degraded in both cases—degradation in activated sewage sludge was completed within 4 days and in the mixed strain consortium in 16 days. [OMIM][Cl] was degraded by activated sewage sludge at almost 60 % by the end of the test which corresponds theoretically to a degradation of the octyl side chain without degradation of the imidazolium ring. Nevertheless, [OMIM][Cl] could not be classified as readily biodegradable on the basis of this result. Biodegradation of [OMIM][Cl] by the mixed strain consortium occurred slower and reached slightly above 30 % on the twenty-eighth day of the test, most probably this also involved the biodegradation of the side chain. The differences in the degradation rates are possibly due to the different cell densities in the consortium and activated sewage sludge as well as different microbial composition. The dry mass cannot be used as a direct comparison, as the dry mass of sewage is partially comprised of a mineral fraction, extracellular polymeric substance as well as other organisms like fungi and protozoa. Therefore, to obtain a comparison of the various cultures, the cell density was determined by counting the colony-forming units. When normalized for cell density (Fig. 3b), the rate of biodegradation is much higher in case of mixed strain consortium. This difference in rates is of course only an indicative result, since the cell density of sewage is several orders of magnitude higher (and therefore, the % degradation/CFU will always be lower). However, it highlights the potential that such mixed strain consortia might have.

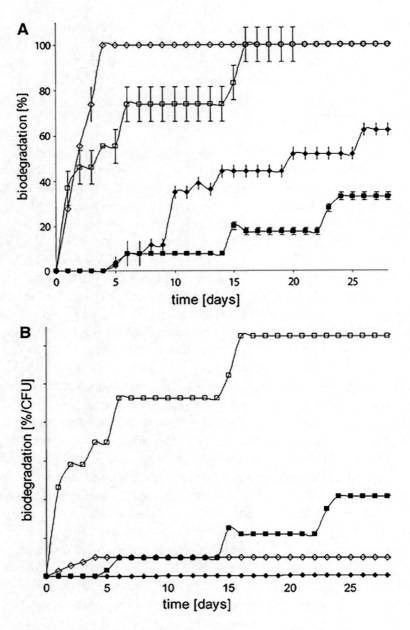

Figure 3. Biodegradation (a) and biodegradation normalized for cell density (b) of [OMIM][Cl] (closed symbols) and sodium glutamate (open symbols) by activated sewage sludge (diamonds) and consortium of nine isolated strains (squares).

5.4 CONCLUSION

Docherty et al. previously reported on changes in microbial genetic patterns of organisms cultivated in media containing ILs, suggesting that enrichment of certain species occurred (Docherty et al. 2007). We decided to follow this lead and isolated bacterial strains from activated sewage sludge subjected to selective pressure of 1-methyl-3-octylimidazolium chloride.

Nine bacterial isolates were identified. Three of them belonged to *Flavobacteriaceae* family, two to family *Micrococcaceae* and the remaining strains to the families: *Shewanellacae*, *Moraxellaceae*, *Microbacteriaceae*, *Nocardiaceae*. All strains grew well in 0.2 mM [OMIM][Cl] and were inhibited in 40 % on average in 2 mM solutions. Any higher concentrations inhibited growth almost completely.

Axenic cultures of single isolates were proven to be rather inefficient in degrading [OMIM][Cl] and application of a consortium composed of all mixed strains resulted in 30 % ultimate degradation. In the same conditions, activated sewage sludge organisms degraded almost 60 % of the tested compound which corresponds to full degradation of alkyl substituents. The lower result obtained for the mixed strain consortium might have been a result of lower cell density in these samples. A higher degradation rate was observed in case of mixed strains when results were normalized for cell density in consortium. It is possible that not all of isolates involved in ILs metabolism were selected or that isolated microorganisms were a part of more complex consortium connected by symbiotic relations with other organisms and that, in their presence, even higher metabolic capacities could have been achieved.

We have recently shown that adapted microbial communities can degrade [OMIM][Cl] faster and can withstand higher concentrations of that IL without an inhibitory effect compared with non-adapted communities (Markiewicz et al. 2011). Therefore, more research is needed in order to uncover species involved in [OMIM][Cl] degradation and to examine the feasibility of using adapted single strains or consortia in enhancing degradative abilities of indigenous microorganisms. Obtaining high cell density inocula of strains isolated here and comparing them with activated sludge of the same cell density would unequivocally confirm their better performance. Additionally, simulation tests in high biomass content systems (e.g., OECD tests 303 OECD 2001) in combination with augmentation of freshly sampled activated sludge with nine isolates could help to prove the technological viability of this approach.

REFERENCES

1. Abrusci C, Palomar J, Pablos JL, Rodriguez F, Catalina F (2010) Efficient biodegradation of common ionic liquids by Sphingomonas paucimobilis bacterium. Green Chem 3:709–717
2. Adebusoye SA, Ilori MO, Amund OO, Teniola OD, Olatope SO (2007) Microbial degradation of petroleum hydrocarbons in a polluted tropical stream. World J Microbiol Biotechnol 23(8):1149–1159
3. Blazevic DJ (1976) Current taxonomy and identification of nonfermentative gram negative bacilli. Hum Pathol 7(3):265–275
4. Calabrese EJ (2005) Paradigm lost, paradigm found: the re-emergence of hormesis as a fundamental dose response model in the toxicological sciences. Environ Pollut 138:378–411
5. Coleman D, Gathergood N (2010) Biodegradation studies of ionic liquids. Chem Soc Rev 39:600–637
6. Docherty KM, Dixon JK, Jr CFK (2007) Biodegradability of imidazolium and pyridinium ionic liquids by an activated sludge microbial community. Biodegradation 18:481–493
7. Gathergood N, Scammells PJ, Garcia MT (2006) Biodegradable ionic liquids Part III: the first readily biodegradable ionic liquids. Green Chem 8:156–160
8. Grady LCPJ (1985) Biodegradation: its measurement and microbiological basis. Biotechnol Bioeng 27:660–674
9. Heipieper HJ, deBont JAM (1994) Adaptation of Pseudomonas putida S12 to ethanol and toluene at the level of fatty acid composition of membranes. Appl Environ Microbiol 60(12):4440–4444
10. Isken S, deBont JAM (1998) Bacteria tolerant to organic solvents. Extremophiles 2:229–238
11. Jiang T, Brym MJC, Dubé G, Lasia A, Brisard GM (2006) Electrodeposition of aluminium from ionic liquids: part II—studies on the electrodeposition of aluminum from aluminum chloride (AICl3)—trimethylphenylammonium chloride (TMPAC) ionic liquids. Surf Coat Technol 201:10–18
12. Khehra MS, Saini HS, Sharma DK, Chadha BS, Chimni SS (2005) Comparative studies on potential of consortium and constituent pure bacterial isolates to decolorize azo dyes. Water Res 39(20):5135–5141
13. Kragl U, Eckstein M, Kaftzik N (2002) Enzyme catalysis in ionic liquids. Curr Opin Biotechnol 13:565–571

14. Krossing I, Slattery JM, Daguenet C, Dyson PJ, Oleinikova A, Weingaertner H (2006) Why are ionic liquids liquid? A simple explanation based on lattice and solvation energies. J Am Chem Soc 128:13427–13434
15. Letunic I, Bork P (2007) Interactive tree of life (iTOL): an online tool for phylogenetic tree display and annotation. Bioinformatics 23(1):127–128
16. Limbergen HV, Top EM, Verstraete W (1998) Bioaugmentation in activated sludge: current features and future perspectives. Appl Microbiol Biotechnol 50:16–23
17. Łuczak J, Jungnickel C, Łącka I, Stolte S, Hupka J (2010) Antimicrobial and surface activity of 1-alkyl-3-methylimidazolium derivatives. Green Chem 12:593–601
18. Markiewicz M, Jungnickel C, Markowska A, Szczepaniak U, Paszkiewicz M, Hupka J (2009) 1-Methyl-3-octylimidazolium chloride—sorption and primary biodegradation analysis in activated sewage sludge. Molecules 14:4396–4405
19. Markiewicz M, Stolte S, Lustig Z, Łuczak J, Skup M, Hupka J, Jungnickel C (2011) Influence of microbial adaption and supplementation of nutrients on the biodegradation of ionic liquids in sewage sludge treatment processes. J Hazard Mater 195(15):378–382
20. Modelli A, Sali A, Galletti P, Samori C (2008) Biodegradation of oxygenated and non-oxygenated imidazolium-based ionic liquids in soil. Chemosphere 73(8):1322–1327
21. NCBI (2011) GenBank. From http://www.ncbi.nlm.nih.gov/genbank/
22. OECD (1992) OECD guideline for testing of chemicals 301—ready biodegradability
23. OECD (2001) Simulation test—aerobic sewage treatment
24. Pham TPT, Cho C-W, Jeon C-O, Chung Y-J, Lee M-W, Yun Y-S (2009) Identification of metabolites involved in the biodegradation of ionic liquid 1-butyl-3-methylpyridinium bromide by activated sludge microorganisms. Environ Sci Technol 43:516–521
25. Pieper DH, Reineke W (2000) Engineering bacteria for bioremediation. Curr Opin Biotechnol 11:262–270
26. Plechkova NV, Seddon KR (2008) Applications of ionic liquids in the chemical industry. Chem Soc Rev 37:123–150
27. Romero A, Santos A, Tojo J, Rodrıguez A (2008) Toxicity and biodegradability of imidazolium ionic liquids. J Hazard Mater 151:268–273
28. Schink B (2002) Synergistic interactions in the microbial world. Antonie Van Leeuwenhoek 81(1–4):257–261
29. Stolte S, Abdulkarim S, Arning J, Blomeyer-Nienstedt A-K, Bottin-Weber U, Matzke M, Ranke J, Jastorff B, Thöming J (2008) Primary biodegradation of ionic liquid cations, identification of degradation products of 1-methyl-3-octyl-imidazolium chloride and electrochemical waste water treatment of poorly biodegradable compounds. Green Chem 10:214–224
30. Takenaka S, Tonoki T, Taira K, Murakami S, Aoki K (2007) Adaptation of Pseudomonas sp. strain 7–6 to quaternary ammonium compounds and their degradation via dual pathways. Appl Environ Microbiol 73(6):1797–1802
31. Volkering F, Breure AM, Andel JGv, Rulkens WH (1995) Influence of nonionic surfactants on bioavailability and biodegradation of polycyclic aromatic hydrocarbons. Appl Environ Microbiol 61(5):1699–1705

32. Weber FJ, deBont JAM (1996) Adaptation mechanisms of microorganisms to the toxic effects of organic solvents on membranes. Biochim Biophys Acta 1286:225–245
33. Zwieten LV, Kennedy IR (1995) Rapid degradation of atrazine by Rhodococcus sp. NI86/21 and by an atrazine-perfused soil. J Agric Food Chem 43(5):1377–1382

PART III

ASSESSMENT OF SEWAGE SLUDGE HAZARD, PRE- AND POST-TREATMENT

CHAPTER 6

Seeking Potential Anomalous Levels of Exposure to PCDD/Fs and PCBs Through Sewage Sludge Characterization

ELENA CRISTINA RADA, MARCO SCHIAVON, AND MARCO RAGAZZI

6.1 INTRODUCTION

Within the large family of Persistent Organic Pollutants (POPs), organochlorine substances like polychlorinated dibenzo-p-dioxins (PCDDs), polychlorinated dibenzofurans (PCDFs) and polychlorinated biphenyls (PCBs) are considered the most toxic and widespread compounds [1]. PCDD/Fs are commonly named dioxin, and are composed of a total number of 210 congeners, who's the most toxic for humans and animals could be grouped in 17 congeners: 7 for PCDDs and 10 for PCDFs [2]. PCBs are known to be composed of a total of 209 congeners, but the most toxic ones are the co-planar PCBs, which are 12, and present a toxic behavior similar to PCDD/Fs; for this reason, these congeners are named dioxin-like PCBs. The exposure to such compounds has been object of important concern in the last decades [3], especially in the light of their health effects. Acute

---

© 2013 Rada EC, et al. J Bioremed Biodeg 4:210. doi: 10.4172/2155-6199.1000210. Creative Commons Attribution license (http://creativecommons.org/licenses/by/3.0/). Used with the authors' permission.

toxicity does not represent the most important hazard; the primary risk is related to chronic exposures at lower concentrations. 2,3,7,8-TCDD, the most toxic PCDD/F congener, was classified as carcinogenic to humans by the International Agency for the Research on Cancer (IARC). This compound is the most known cancer promoter, since increased risks for lung cancer, soft-tissue sarcoma, non-Hodgkin lymphoma and other malignant neoplasms were reported in several cohort studies [4,5]. There is also clear evidence that PCBs cause cancer in animals, especially liver and thyroid neoplasms [6,7]. In addition, the results of a number of epidemiological studies raise concerns for the potential carcinogenicity of PCBs on humans [7]. For these reasons, the IARC classified PCBs in the Group 2A, as potential carcinogenic to humans [8].

More than the 90% of the average dioxin daily intake is estimated deriving from food consumption, primarily dairy products, followed by cereals and vegetables, meat and fish [9,10]. In fact, POPs have lipophilic properties and bioaccumulation represents the prevailing way of contamination of the food chain: atmospheric deposition of POPs coming from various sources (e.g. waste treatment plants, production of chemicals, metal industry) firstly contaminates soil, hay, vegetables and fruit [11]; contaminated soil and hay transfer the accumulated POPs to the adipose tissues of the cattle, and following the consumption of meat and dairy products to the humans, whilst contaminated vegetables and fruit may transfer their POP content to the cattle, but also directly to humans. Thus, it is clear the importance of methods for preventing the alteration of soils by POPs, avoiding remediation interventions. The absorption of PCDD/Fs and PCBs from the diet can result higher than 80% in situations of elevated POP intake [12]. The absorption is followed by excretion from the body, which does not depend on fluctuations of the concentration in the diet, but only on the concentration in the body: in fact, the faecal elimination occurs continuously at a rate that depends on the blood concentration [12]. The body, thus, seems to be able to equalize the effects of peak intakes of POPs.

On European scale, it was estimated that steel making plants would constitute the most important source of PCDD/Fs [13]. The metal sector is also an important PCB source, as demonstrated by a study on scrap metal recycling plants, whose wastewater showed a maximal dioxin-like PCB concentration of 3.61 ngI-TEQ L-1 [14].

After entering the human body and leaving the organism by excretion, POPs reach the wastewater treatment plants (WWTPs) and concentrate in sewage sludge. Sludge contamination is then object of great concern, if the reuse of sewage sludge in agriculture is an option [15]. Two important studies focused on the mass balance of POPs throughout the human body [16,12], showing the significant role of bioaccumulation.

The presence of organic [17,18] and inorganic [19,20] hazardous substances in sewage sludge has been object of investigation in several studies. Sewage sludge has also been studied as potential source of energy [21-26], and as raw material for conventional [27,28] and nonconventional products [29,30]. In addition to these applications, sewage sludge can be also used as source of information to seek potential anomalous levels of exposure of a population to POPs: indeed, the dominant exposure route of POPs released by the most important sources (e.g. steel making plants) is the emission into the atmosphere, the atmospheric deposition to farmlands and pastures, the consequent contamination of the food produced, the intake by humans, the excretion, the transportation to WWTPs and the POP concentration in sewage sludge; thus, analyses on the POP content of sludge samples offer an alternative, inexpensive and technically simple approach to assess the existence of critical situations of exposure to POPs. Several studies focused on the ambient air and deposition monitoring in areas where steel making plants were present and highlighted their evident influence in terms of PCDD/F contribution in the surroundings [31-33]. The present study is then intended to propose a methodology, in order to detect anomalies throughout the food chain, as a consequence of the release of POPs in air from significant sources and their subsequent deposition to farmlands and pastures.

This study focuses on a steel making plant located in a West-East oriented Alpine valley in the North of Italy. Three domestic WWTPs located outside the area of influence of the plant, were chosen as background reference for assessing the content of POPs in their sewage sludge. Another domestic WWTP, which receives the wastewater from the villages located in the vicinity of the steel making plant, was chosen as representative of the population exposed to the intake of POPs released by the mill. One sewage sludge sample was taken from each WWTP. The samples were analyzed and the results are discussed, in order to understand the potentialities of this methodology in assessing the presence of critical levels of exposure to POPs. In addition, a sensitivity analysis of the method is presented and the results are interpreted by moving from the considerations presented [12], and from the findings of a previous study, which reported that a permanent exposure to normal levels of PCDD/Fs, following a higher exposure in the past can produce a PCDD/F excretion that is twice higher than the intake [34]. Considered the limited number of samples, the application of this approach to a specific case study has the purpose of explaining how this methodology can be applied in different contexts, with a sufficient number of samples to perform a statistical analysis and better interpret the results.

6.2 EXPERIMENTAL SECTION

To highlight the effect of PCDD/F and PCB emissions from the steel making plant on the food chain and to reconstruct the fate of these compounds into the environment, an analysis of four sewage sludge samples from four different WWTPs was performed. One of them was chosen as reference to assess the direct impact of the emissions from the steel making plant on the food chain of the population that lives inside the area of influence of the mill. The choice of this WWTP started from previous dispersion simulations of PM emitted by the plant, whose results allowed the detection of the population potentially exposed to contamination, through direct inhalation and assumption via food intake after deposition [35]. The region surrounding the plant hosts several cultivated lands, cattle and dairy farms. The consumption of locally produced food was assumed equal to 10% of the total diet of the local population, according to a previous study [36].

The steel plant-influenced WWTP located inside the area of influence of the steel making plant, was finally chosen from those plants whose catchment area includes the villages directly exposed to the emissions, according to the dispersion simulations [35]. The input for this WWTP is composed for 75% by the wastewater from the potentially exposed population (about 12,670 inhabitants); the remaining 25% is composed by wastewater from other residential unexposed populations [37]; moreover, the WWT line receives liquid streams from the thermal drying of provincial sewage sludge that can be supposed poor of dioxin. In fact, the liquid streams are dominated by the liquid phase, unlike domestic wastewater that has a higher content of suspended solids (or organic matter): indeed, a positive association between dioxin levels and suspended solids was found in a previous study, which supports the theory that dioxin-like compounds will partition almost exclusively onto the organic matter in sludge in preference to water [38].

As a term of comparison, three more WWTPs were chosen as representative of the population living outside the study area (about 39,690 inhabitants), hence providing background information on the PCDD/F and PCB levels in the food chain. Similarly, these three plants were chosen so that their catchment areas include villages located outside the area of influence of the plant and far from significant emitters. The consultation of the local emission inventory excluded the presence of other important sources of POPs. Locations of the WWTPs and their distances from the steel making plant are presented in Figure 1.

All the WWTPs here considered are equipped with an oxidation stage and a secondary sedimentation tank; the sludge is then conditioned with the addition of polyelectrolytes and is sent to mechanical dehydration. The choice of taking sludge samples instead of wastewater samples is related to the fact that POPs are more concentrated in sludge than in water. As a matter of fact, the activated sludge treatment that is applied to WWTPs removes hydrophobic compounds like PCDD/Fs and PCBs from the wastewater by sorption to sludge [39]. The four sewage sludge samples (sample volume of 2 l each), one for the steel plant-influenced WWTP and three for the background WWTPs, were taken on the 12.12.2011, between 9 am and 11 am. The choice of this period of the year is crucial and is due to the need of avoiding contributions from tourist peaks in the region, which usually occur during summer or during Christmas holidays. In fact, by taking the sludge samples during the low season, only the contribution of the resident population is counted. If sampling had been carried out during a tourist period, the results would have shown the diluting effect of a differently exposed foreign population. Since POP levels in sewage sludge change slowly, due to the slowness of the process of accumulation in the food chain, the number of samples taken at each plant is not influent on the characterization of the population exposure in a specific period (tourist periods excluded). On the other hand, primary importance should be given to the choice of the WWTPs that must be representative of the exposed and the unexposed population and must clearly distinguish between them.

The four sludge samples were taken before the chemical conditioning with polyelectrolytes. Since polyelectrolytes are synthesis products, the possibility exists that such substances may contain dioxin compounds, and thus, they may increase the dioxin content of the unconditioned sludge. The choice of sampling sewage sludge before conditioning gives the certainty that no additional sources of dioxin (other than anthropic input) influenced the sludge samplings. Furthermore, cross-contamination was prevented by the adoption of glass vases as containers for the samples. Hence, the only possible source of dioxin and dioxin-like compounds could be represented by the indoor air, whose quality is kept under control and possible dioxin levels are anyway negligible if compared with the concentration in the samples.

The moisture of the samples was detected by calculation of the dry residual mass after evaporation at 105°C, according to the EN 14346:2007 method [40]. The content of PCDD/Fs was measured in accordance with the EPA 1613 method: the sample was extracted in a Soxhlet/Dean-Stark (SDS) extractor; the extract was concentrated for cleanup and 37Cl4-labeled 2,3,7,8-TCDD was added to measure the efficiency of the cleanup process; the extract was then concentrated to near dryness and injected into the gas chromatograph (GC) after addition of internal standards; the analytes were detected by high resolution mass spectrometry (HRMS); isotope dilution and internal standard techniques were used for quantitative analyses, according to the congeners to be measured [41]. The content of dioxin-like PCBs in the sludge samples was determined following the EPA 1668B methodology: the sample was extracted by an SDS extractor and the extract was cleaned up by back-extraction with sulphuric acid; the extract was concentrated to 20 μL and charged with internal standards; the analytes were detected by HRMS and quantitatively measured by isotope dilution [42].

Seeking Potential Anomalous Levels of Exposure to PCDD/Fs and PCBs

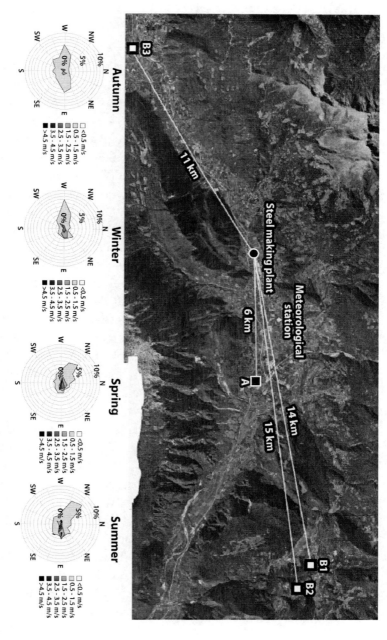

Figure 1. Seasonal wind roses for the case study and locations of the steel making plant (blue circle), the meteorological station (light blue rhombus), the steel plant influenced WWTP (red square), the three background WWTPs (green squares) and their respective distances from the steel making plant.

6.3 RESULTS AND DISCUSSIONS

The results of the analysis on the four sludge samples from the WWTP taken as reference for the population potentially exposed to contamination (A) and for the three background WWTPs (B1, B2 and B3) are presented in Table 1 and 2, for PCDD/Fs and PCBs, respectively. For some PCDD/F congeners (e.g. 2,3,7,8-TCDD and 1,2,3,7,8,9-HxCDF), the concentration measured in sewage sludge is lower than the instrumental detection limit (DL) for all the WWTPs considered. Conventionally, concentrations below the DL are assumed as half the DL itself (Table 1 and 2).

As the results show, the total PCDD/F concentrations measured at the steel plant-influenced WWTP (A) are slightly higher than those measured elsewhere, since the congeners 1,2,3,4,6,7,8-HpCDD and OCDD are predominant. Moreover, such compounds are two of the most important congeners that characterize the emissions from an electric arc furnace [43], which is the technology adopted by the steel making plant under investigation. On the other hand, in terms of total WHO-TEQ concentrations, the A plant shows the lowest dioxin content with respect to the others. This behavior is due to the fact that the toxicity of the dominant congeners (1,2,3,4,6,7,8-HpCDD and OCDD) is lower than the others (Table 1). Considering that the WHO-TEQ concentrations are related to the toxicity for humans, no anomalies in PCDD/Fs were detected in the food chain for the potentially exposed population.

These results are in agreement with the fact that the PCDD/F deposition in the surroundings of the steel making plant, in terms of WHO-TEQ values, is substantially low and can be considered similar to that normally found in rural areas, as demonstrated by a monitoring campaign carried out between 2010 and 2011 [44]. The use of wood as a source for domestic heating, which is typical of mountainous regions where this material is abundant, may have had a certain influence in making the results of the four samples comparable. The PCDD/F concentrations found in the samples are within the range 1.2–15.3 ng WHO-TEQ kg^{-1} DM measured in a previous study on Australian WWTPs [45], and are lower than the average concentrations found in two German WWTPs, expressed in terms of International Toxic Equivalency (25 and 10 ng I-TEQ kg^{-1}DM) [46].

Seeking Potential Anomalous Levels of Exposure to PCDD/Fs and PCBs 111

Table 1. PCDD/F concentrations and respective WHO-TEQ concentrations in the sludge samples for the WWTP taken as reference for the population exposed to POP contamination (A) and for the background WWTPs (B1, B2 and B3); concentrations below the DL are assumed equal to half the DL itself for the calculation of the total concentrations.

Congener	WHO-TEF	DL [ng kg⁻¹ DM]	Concentration [ng kg⁻¹ DM]				WHO-TEQ Concentration [ng WHO-TEQ kg⁻¹ DM]			
			A	B1	B2	B3	A	B1	B2	B3
2,3,7,8-TCDD	1	0.1	<DL	<DL	<DL	<DL	0.050	0.050	0.050	0.050
1,2,3,7,8-PCDD	1	0.5	<DL	0.60	<DL	<DL	0.250	0.600	0.250	0.250
1,2,3,4,7,8-HxCDD	0.1	0.5	<DL	<DL	0.60	<DL	0.025	0.025	0.060	0.025
1,2,3,6,7,8-HxCDD	0.1	0.5	2.60	1.10	4.00	<DL	0.260	0.110	0.400	0.025
1,2,3,7,8,9-HxCDD	0.1	0.5	<DL	<DL	1.20	1.70	0.025	0.025	0.120	0.170
1,2,3,4,6,7,8-HpCDD	0.01	0.5	57.00	31.00	46.00	48.00	0.570	0.310	0.460	0.480
OCDD	0.0001	0.5	466.00	323.00	355.00	412.00	0.047	0.032	0.036	0.041
2,3,7,8-TCDF	0.1	0.5	4.10	1.60	2.40	3.30	0.410	0.160	0.240	0.330
1,2,3,7,8-PCDF	0.05	0.5	0.80	0.60	0.60	0.90	0.040	0.030	0.030	0.045
2,3,4,7,8-PCDF	0.5	0.5	<DL	0.50	1.70	1.70	0.125	0.250	0.850	0.850
1,2,3,4,7,8-HxCDF	0.1	0.5	<DL	0.70	1.40	1.60	0.025	0.070	0.140	0.160
1,2,3,6,7,8-HxCDF	0.1	0.5	0.80	1.70	0.90	1.00	0.080	0.170	0.090	0.100
2,3,4,6,7,8-HxCDF	0.1	0.5	0.60	1.60	2.00	<DL	0.060	0.160	0.200	0.025
1,2,3,7,8,9-HxCDF	0.1	0.5	<DL	<DL	<DL	<DL	0.025	0.025	0.025	0.025
1,2,3,4,6,7,8-HpCDF	0.01	0.5	17.90	16.70	6.20	16.40	0.179	0.167	0.062	0.164
1,2,3,4,7,8,9-HpCDF	0.01	0.5	<DL	0.50	0.60	<DL	0.003	0.005	0.006	0.003
OCDF	0.0001	0.5	44.00	49.00	59.00	52.00	0.004	0.005	0.006	0.005
Total	-	-	595.60	429.40	482.15	540.15	2.178	2.194	3.024	2.748

Table 2. PCB concentrations and respective WHO-TEQ concentrations in sludge samples for the WWTP taken as reference for the population exposed to POP contamination (A) and for the background WWTPs (B1, B2 and B3); concentrations below the DL are assumed equal to half the DL itself for the calculation of the total concentrations.

Congener	WHO-TEF	DL [ng kg$^{-1}_{DM}$]	Concentration				WHO-TEQ Concentration			
			[ng kg$^{-1}_{DM}$]				[ng WHO-TEQ kg$^{-1}_{DM}$]			
			A	B1	B2	B3	A	B1	B2	B3
PCB 77	0.0001	1	492.0	71.0	172.0	191.0	0.0492	0.0071	0.0172	0.0191
PCB 81	0.0003	1	19.6	<DL	1.4	1.9	0.0059	0.0002	0.0004	0.0006
PCB 105	0.00003	1	1890.0	233.0	650.0	700.0	0.0567	0.0070	0.0195	0.0210
PCB 114	0.00003	1	162.0	<DL	48.0	59.0	0.0049	0	0.0014	0.0018
PCB 118	0.00003	1	5210.0	1290.0	1800.0	1800.0	0.1563	0.0387	0.0540	0.0540
PCB 123	0.00003	1	403.0	35.0	136.0	126.0	0.0121	0.0011	0.0041	0.0038
PCB 126	0.1	1	5.2	<DL	1.4	7.1	0.5200	0.0500	0.1400	0.7100
PCB 156	0.00003	1	600.0	151.0	343.0	343.0	0.0180	0.0045	0.0103	0.0103
PCB 157	0.00003	1	99.0	32.0	67.0	52.0	0.0030	0.0010	0.0020	0.0016
PCB 167	0.00003	1	141.0	55.0	97.0	114.0	0.0042	0.0017	0.0029	0.0034
PCB 169	0.03	1	7.0	2.1	2.8	6.4	0.2100	0.0630	0.0840	0.1920
PCB 189	0.00003	1	53.0	14.0	20.0	18.8	0.0016	0.0004	0.0006	0.0006
Total	-	-	9081.8	1884.6	3338.6	3419.2	1.0418	0.1746	0.3365	1.0181

Seeking Potential Anomalous Levels of Exposure to PCDD/Fs and PCBs   113

With regard to PCB concentrations, some considerations that slightly differ from the case of PCDD/Fs can be made: the total concentration at the A plant is between 2.7 and 4.8-time higher than the other plants; PCBs are typical compounds emitted by steel making plants [14], and this can explain such a finding. The same sludge sample shows a WHO-TEQ concentration that is higher than the B1 and B2 samples, but is similar to the B3 sample, where the concentration of the most toxic congener (PCB 126) is also higher. The different urbanization, compared to the B1 and B2 plants, can explain this aspect, since the B3 case is characterized by a moderate presence of small industrial activities.

Some considerations about the sensitivity of the methodology can also be expressed. If considering that 75% of the stream is composed by wastewater from the exposed population, the ratio between the POP concentrations visible in the sewage sludge after and before the increase of exposure (ΔPOP_{sludge}) can be calculated by the following equation:

$$\Delta POP_{sludge} = 0.75\Delta POP_{feces} + 0.25\Delta POP_{other} \qquad (1)$$

Where ΔPOP_{feces} is the ratio between the POP concentrations occurring in the feces after and before the exposure increase and ΔPOP_{other} is the ratio between the POP concentrations (after and before the exposure increase) in the sludge from the unexposed population whose wastewater is collected at plant A.

In conditions of permanent exposure, ΔPOP_{feces} is considered equal to the ratio between the intakes by the human body after and before the exposure increase (ΔPOP_{intake}) [12], which is the sum of the ratio between the POP concentrations in the locally produced food ($\Delta POP_{food,loc}$), the ratio of the POP concentrations in the non-locally produced food ($\Delta POP_{food,nloc}$) and in air (ΔPOP_{air}) after and before the exposure increase:

$$\Delta POP_{feces} = \Delta POP_{intake} = C_{food,loc}I_{POP,food}$$
$$\Delta POP_{food,loc} + C_{food,loc}I_{POP,food}\Delta POP_{food,loc} + = I_{POP,food}\Delta POP_{air} \qquad (2)$$

Where $C_{food,loc}$ is the percentage of the consumption of locally produced food on the total diet, $C_{food,nloc}$ is the percentage of the consumption of non-locally produced food, $I_{POP,food}$ is the percentage of POPs taken in by food consumption, and the remaining part is attributed to the intake of POPs by inhalation ($I_{POP,inhal}$). $C_{food,loc}$ is assumed equal to 10%, according to Cernuschi [36]; consequently, $C_{food,nloc}$ is equal to 90%; following the findings of [9], $I_{POP,food}$ is assumed equal to 90%, and consequently, $I_{POP,inhal}$ is equal to 10%. If considering ΔPOP_{intake} only addressed to the consumption of local food, $\Delta POP_{food,nloc}$ and ΔPOP_{air} are equal to 1. Thus, by applying (2) an increase of concentration of 100 times in the locally produced food ($\Delta POP_{food,loc}=100$) would result in an increase of 9.91 times in the POP content in feces ($\Delta POP_{feces}=9.91$). If assuming the POP concentration in the sludge from the unexposed population whose wastewater is collected at plant A as unvaried ($\Delta POP_{other}=1$), by applying (1), a 7.7-time higher concentration should be observed in the sewage sludge at the steel plant-influenced WWTP.

In this specific case, the assumption of permanent exposure cannot be made, since in 2009, the steel making plant significantly modified the off-gas treatment line in order to decrease its emissions. In this case, the sensitivity of the method can be assessed by considering that, in conditions of permanent exposure to normal levels following a higher exposure in the past, the PCDD/F excretion can result twice higher than the intake [34]. In this case, $\Delta POP_{feces}=2\Delta POP_{intake}$, and if considering a past concentration in the locally produced food that was 100 higher than the background ($\Delta POP_{food,loc}=100$), ΔPOP_{sludge} would be approximately equal to 15.

It should be taken into account that the 2.7-4.8 time higher PCB concentrations found at the A plant can be also explained by the higher exposure that occurred in the past.

On the other hand, the method is not able to detect acute episodes of exposure, since as previously stated, in such situations, even only 20% of the assumed POPs can be excreted [12]: in this case, $\Delta POP_{feces}=0.2\,\Delta POP_{in}$ and ΔPOP_{sludge} would become comparable with that one achievable in permanent conditions, if the concentration in the locally produced food is 12-time higher than the background, thus giving misleading results.

Given that the contribution of food consumption in the intake of POPs is dominant and assuming acceptable, the daily intake of the population living in the areas B1, B2 and B3 (for the absence of significant sources of POPs), the fact that the concentrations of POPs in the sewage sludge at the A plant are similar to those found in the unexposed areas allows excluding the absence of a dioxin and dioxinlike emergency in the area of the steel making plant, even if the presence of the steel making plant is visible by the detection of some congeners characteristic of the emissions from electric arc furnaces. The positive results can be attributed also to the technology adopted, as the arc furnace has a less impact in terms of dioxin emissions with respect to other technologies [13].

Possible wastewater contaminations can derive from surface runoff, but this contribution was found to be limited to about 10% in a previous work [46]. Laundry washing is an additional source of PCDD/Fs, and its contribution can be comparable with the one of domestic wastewater [46]. However, if considering the habits unchanged over the years, the contribution of laundry washing can be considered constant; thus, a hypothetical 100-time higher dietary exposure would be anyway visible in terms of POP concentration in sludge.

6.4 CONCLUSION

A novel, inexpensive and technically simple methodology to detect anomalies in the exposure to POPs throughout the food chain was presented. This methodology, suitable also for preventing soil contamination, was applied to the case of a steel plant and the residential population living in the surroundings, but the approach can be applied to any case where a population is exposed to contamination by POPs. Given the limited number of samples, the application to the present case study has the purpose of better explaining how the methodology works, without providing definitive results on the exposure assessment of the target population. For future applications and for a better reliability and solidity of the method, increasing the number of samples at every WWTP is necessary. Three samplings under the same conditions at each site will allow a statistical analysis of the results.

In spite of the limited number of samples taken in this study, no anomalies in PCDD/Fs were detected for the sludge samples representative of the potentially exposed population, giving credit to the favorable results obtained during a previous deposition monitoring campaign. In terms of PCBs, the WWTP taken as reference for the exposed population showed a total concentration between 2.7 and 4.8-time higher than the other plants; in terms of WHO-TEQ concentrations, only slightly higher PCB levels were found for the WWTP representative of the exposed population, even though the most toxic congener (PCB 126) presented a higher concentration at the B3 plant. In order to understand the meaning of the 2.7-4.8 time higher PCB sludge concentrations found at the A plant, it must be remarked that in 2009 the steel making plant adopted the Best Available Technologies for the air pollution control system, and the levels nowadays measured can still incorporate the effects of the higher exposure occurred in the past. However, considering acceptable the daily intake from the diet of the population living in the areas not affected by the steel plant, the absence of a dioxin and dioxin-like emergency in the area of the steel plant can be deducted by the similar concentrations of PCDD/Fs and PCBs in the sewage sludge samples.

The method is capable of well assessing situations of long periods of exposure of the population to POP levels that are higher than the background, even if difficulties in the interpretation of the results can occur in presence of acute episodes.

REFERENCES

1. Rada EC, Ragazzi M, Marconi M, Chistè A, Schiavon M, et al. (2010) Dioxin and waste sector role: An example. Proceedings of XIII International Waste Management and Landfill Symposium, Margherita di Pula S (Cagliari), Sardinia. Italy.
2. Rada EC, Ragazzi M, Panaitescu V, Apostol T (2006) The role of bio-mechanical treatments of waste in the dioxin emission inventories. Chemosphere 62: 404-410.
3. Rada EC, Ragazzi M (2008) Critical analysis of PCDD/F emissions from anaerobic digestion. Water Sci Technol 58: 1721-1725.
4. Silbergeld EK, Gasiewicz TA (1989) Dioxins and the Ah receptor. Am J Ind Med 16: 455-474.
5. (1997) IARC working group on the evaluation of carcinogenic risks to humans: Polychlorinated dibenzo-para-dioxins and polychlorinated dibenzofurans, Lyon, France. IARC Monogr Eval Carcinog Risks Hum 69: 1-631.
6. Usydus Z, Szlinder-Richert J, Polak-Juszczak L, Komar K, Adamczyk M, et al. (2009) Fish products available in Polish market--Assessment of the nutritive value and human exposure to dioxins and other contaminants. Chemosphere 74: 1420-1428.
7. USEPA (2013) Health effects of PCBs.

8. (1987) Overall evaluations of carcinogenicity: An updating of IARC Monographs volumes 1 to 42. IARC Monogr Eval Carcinog Risks Hum 1-440.
9. Eduljee GH, Gair AJ (1996) Validation of a methodology for modelling PCDD and PCDF intake via the foodchain. Sci Total Environ 187: 211-229.
10. Fürst P, Beck H, Theelen RMC (1992) Assessment of human intake of PCDDs and PCDFs from different environmental sources. Toxic Subst J 12: 133-150.
11. Liem AK, Fürst P, Rappe C (2000) Exposure of populations to dioxins and related compounds. Food Addit Contam 17: 241-259.
12. Moser GA, McLachlan MS (2001) The influence of dietary concentration on the absorption and excretion of persistent lipophilic organic pollutants in the human intestinal tract. Chemosphere 45: 201-211.
13. Quass U, Fermann M, Bröker G (2004) The European dioxin air emission inventory project--Final results. Chemosphere 54: 1319-1327.
14. Van Ham R, Blondeel M, Baert R (2011) Dioxin, furans and dioxin-like PCBs in wastewater from industry in the Flemish region (Belgium). Organohalogen Compd 73: 162-165.
15. Olofsson U, Brorström-Lundén E, Kylin H, Haglund P (2013) Comprehensive mass flow analysis of Swedish sludge contaminants. Chemosphere 90: 28-35.
16. Juan CY, Thomas GO, Semple KT, Jones KC (1999) An input–output balance study for PCBs in humans. Organohalogen Compd 44: 153-156.
17. Torretta V (2012) PAHs in wastewater: Removal efficiency in a conventional wastewater treatment plant and comparison with model predictions. Environ Technol 33: 851-855.
18. McLachlan MD, Horstmann M, Hinkel M (1996) Polychlorinated dibenzo-p-dioxins and dibenzofurans in sewage sludge: Sources and fate following sludge application to land. Sci Total Environ 185: 109-123.
19. Smith SR (2009) A critical review of the bioavailability and impacts of heavy metals in municipal solid waste composts compared to sewage sludge. Environ Int 35: 142-156.
20. Jamali MK, Kazi TG, Arain MB, Afridi HI, Jalbani N, et al. (2009) Heavy metal accumulation in different varieties of wheat (Triticum aestivum L.) grown in soil amended with domestic sewage sludge. J Hazard Mater 164: 1386-1391.
21. Havukainen J, Zavarauskas K, Denafas G, Luoranen M, Kahiluoto H, et al. (2012) Potential of energy and nutrient recovery from biodegradable waste by co-treatment in Lithuania. Waste Manag Res 30: 181-189.
22. Ragazzi M, Grigoriu M, Rada EC, Malloci E, Natolino F (2010) Health risk assessment from combustion of sewage sludge: A case study with comparison of three sites. In: Recent advances in risk management, assessment and mitigation, Proceedings of International conference on risk management, assessment and mitigation, Bucharest, Romania.
23. Ragazzi M, Rada EC, Ferrentino R (2013) Analysis of real scale experiences of novel sewage sludge treatments in an Italian pilot region. Proceedings of 13th International Conference on Environmental Science and Technology, Athens, Greece.
24. Zhai Y, Wang C, Chen H, Li C, Zeng G, et al. (2013) Digested sewage sludge gasification in supercritical water. Waste Manag Res 31: 393-400.

25. Balgaranova J (2003) Plasma chemical gasification of sewage sludge. Waste Manag Res 21: 38-41.
26. Torretta V, Conti F, Leonardi M, Ruggieri G (2012) Energy recovery from sludge and sustainable development: A Tanzanian case study. Sustainability 4: 2661-2672.
27. Ingelmo F, Canet R, Ibañez MA, Pomarez F, García J (1998) Use of MSW compost, dried sewage sludge and other wastes as partial substitutes for peat and soil. Bioresour Technol 63: 123-129.
28. Tsai W.T. (2012) An analysis of waste management policies on utilizing biosludge as material resources in Taiwan. Sustainability 4: 1879-1887.
29. Tay JH, Chen XG, Jeyaseelan S, Graham N (2001) Optimising the preparation of activated carbon from digested sewage sludge and coconut husk. Chemosphere 44: 45-51.
30. Monsalvo VM, Mohedano AF, Rodriguez JJ (2011) Activated carbons from sewage sludge: Application to aqueous-phase adsorption of 4-chlorophenol. Desalination 227: 377-382.
31. Onofrio M, Spataro R, Botta S (2011) The role of a steel plant in north-west Italy to the local air concentrations of PCDD/Fs. Chemosphere 82: 708-717.
32. Li Y, Wang P, Ding L, Li X, Wang TX, et al. (2010) Atmospheric distribution of polychlorinated dibenzo-p-dioxins, dibenzofurans and dioxin-like polychlorinated biphenyls around a steel plant Area, Northeast China. Chemosphere 79: 253-528.
33. Fang M, Choi SD, Baek SY, Park H, Chang YS (2011) Atmospheric bulk deposition of polychlorinated dibenzo-p-dioxins and dibenzofurans (PCDD/Fs) in the vicinity of an iron and steel making plant. Chemosphere 84: 894-899.
34. Schrey P, Wittsiepe J, Mackrodt P, Selenka F (1998) Human fecal PCDD/F-excretion exceeds the dietary intake. Chemosphere 37: 1825-1831.
35. Rada EC, Ragazzi M, Chistè A, Schiavon M, Tirler W, et al. (2012) A contribution to the evolution of the BAT concept in the sintering plant sector. Proceedings of 9th international symposium of sanitary-environmental engineering, Milano, Italy.
36. Cernuschi S (2003) Municipal waste incineration plant of Trento: Characterization of the presence of toxic trace pollutants in the established area and health risk analysis. Technical Report, Politecnico di Milano, Italy.
37. ADEP, Online database of the wastewater treatment plants of the Autonomous Province of Trento.
38. Telliard WA, McCarty HB, King JR, Hoffman JB (1990) USEPA national sewage sludge survey results for polychlorinated dibenzo-p-dioxins and polychlorinated dibenzofurans. Organohalogen Compd 2: 307-310.
39. Ju JH, Lee IS, Sim WJ, Oh JE, Eun H, et al. (2007) Evaluation of PCDD/Fs and coplanar-PCBs in sewage sludge from wastewater treatment plants in Korea. Organohalogen Compd 69: 1491-1494.
40. UNI EN 14346:2007– Waste characterization–Calculation of dry matter through determination of the dry residual or water content.
41. USEPA, Method 1613-Tetra-through octa-chlorinated dioxins and furans by isotope dilution HRGC/HRMS.
42. USEPA, Method 1668B - Chlorinated biphenyl congeners in water, soil, sediment, biosolids, and tissue by HRGC/HRMS.

43. Zou C, Han J, Fu Hx (2012) Emissions of PCDD/Fs from steel and secondary nonferrous productions. Procedia Environ Sci 16: 279-288.
44. Ragazzi M, Rada EC, Girelli E, Tubino M, Tirler W (2011) Dioxin deposition in the surroundings of a sintering plant. Organohalogen Compd 73: 1920-1923.
45. Clarke B, Porter N, Symons R, Blackbeard J, Ades P, et al. (2008) Dioxin-like compounds in Australian sewage sludge--Review and national survey. Chemosphere 72: 1215-1228.
46. McLachlan MS, Horstmann MX, Hinkel M (1996) Polychlorinated dibenzo-p-dioxins and dibenzofurans in sewage sludge: Sources and fate following sludge application to land. Sci Tot Environ 185: 109-123.

CHAPTER 7

Occurrence and Distribution of Synthetic Organic Substances in Boreal Coniferous Forest Soils Fertilized with Hygienized Municipal Sewage Sludge

RICHARD LINDBERG, KENNETH SAHLÉN, AND MATS TYSKLIND

7.1 INTRODUCTION

According to the Swedish government in 2006, the growth of the Swedish forest should increase 20% the following ten years, by means of fertilization, in order to replace the use of fossil fuels [1]. A similar conclusion was made during the evaluation of the Forest Bill 2007 [2]. In addition, the usage of such fertilizers in forest land should increase. The most essential nutrient for growth is nitrogen, and forest fertilization with nitrogen based fertilizers has been done over a long period [3]. An increase in growth, in the range of 15–20 $m^3 \cdot ha^{-1}$, is possible with a nitrogen dose of 150 $kg \cdot ha^{-1}$. Today, approximately 60,000 ha is fertilized in Sweden each year. On withdrawal of whole trees and rejected tops and branches, a larger nutrient loss is expected via needles, in comparison to the traditional collection of timber and pulpwood. In addition, the losses in growth due to nitrogen

---

© 2013 by the authors; licensee MDPI, Basel, Switzerland. Antibiotics 2013, 2(3), 352-366; doi:10.3390/antibiotics2030352. Creative Commons Attribution license (http://creativecommons.org/licenses/by/3.0/).

deficiency may also follow thinning [4]. Today, wood ash is recommended in order to compensate for the nutrient output following collection of tops and branches [5]. However, the ash lacks nitrogen and may cause growth reductions in less fertile soils [6].

In Sweden, about 240,000 tons of municipal sewage sludge (dry matter, dm) are produced and, according to one of the Swedish environmental goals, at least 60% of phosphorus should be reused as fertilizer [7]. Sewage sludge contains all of the essential nutrients to prevent growth loss following withdrawal of whole trees, and rejected tops and branches. The nitrogen and the organic matter content in sludge contribute to a higher production and will improve the soil's ability to maintain the nutrients. The use of dried granular sludge as a fertilizing agent has shown that growth increases of at least 50% can be obtained [8]. A significant portion of the nitrogen in the sludge is organically bound allowing higher doses, in comparison to nitrogen fertilizers, without leakage of nitrate via ground water. In North America positive effects were noticed for more than 15 years following fertilizing forests with sewage sludge [9]. It has also been shown that if sewage sludge is used as a nitrogen source, carbon emissions caused by the manufacturers of fertilizers will decrease [10].

Municipal sewage sludge does not only consist of nutrients, on the contrary, heavy metals and organic substances that may cause adverse environmental effects are also present. The levels of heavy metal levels are significantly lower in sludge in comparison to wood ash (e.g., only 10% cadmium), and if the nutritional compensation is made with sludge, the total load of metals would be lower. Numerous studies of the effects of heavy metals in the forest ecosystem have been made, and they conclude that it is unlikely that adverse environmental effects may occur in a practical application of forest fertilizing [11]. However, in terms of the organic contaminants, the knowledge regarding their fate in boreal coniferous forests, following fertilization with sewage sludge, is poor. Contaminants associated to municipal sewage sludge include: flame retardants and plasticizers, e.g., polybrominated diphenyl ethers (PBDEs) and polychlorinated biphenyls (PCBs); byproducts, e.g., polyaromatic hydrocarbons (PAHs); antibacterial agents in hygiene products, e.g., triclosan (TCS); and pharmaceutical residues such as fluoroquinolone antibiotics (FQs) and ethinyl estradiol (EE2) [12,13,14,15]. In Sweden, the use of

PCBs was banned in 1972, but during the gradual phasing out, PCB could be present in imported closed systems, e.g., capacitors and transformers, during the following 20 years. From 1995 and forward, PCBs should not exist in any products or systems in Sweden [16]. The regulation of PAHs (European Community) comprises a sales ban if the total concentration of the eight prioritized PAHS is greater than 10 mg·kg^{-1} (1 mg·kg^{-1} for Benzo[a]pyrene) in tires or the oil for the manufacturing of tires [17]. Other sources of PAHs include car exhaust, the wearing down of tire and road materials, and petrol stations. A total ban of PBDEs within the European Union has not yet been initiated, with the exception of penta- and octabrominated diphenyls in 2004, and PBDE #209 in 2006 (not be used in electronic products). The use of TCS has been questioned worldwide and the Swedish Society for Nature Conservation requires a total ban on this substance in any consumer products (e.g., in toothpastes and textiles) due to its environmental hazards [18]. The Swedish Pharmaceutical Industry Association has launched a simple environmental classification (based on information from the pharmaceutical manufacturers) with the goal that all active pharmaceutical ingredients will have environmentally relevant information within five years [19]. Negative environmental impact of EE2 (used in combination birth control pills) is not excluded since ecotoxicological information is missing. FQs (used in treatment of conditions such as urinary tract infection) have been classified by some of the manufacturers as (ciprofloxacin only): may cause a negative environmental impact, potentially persistent, and not readily bioaccumulative.

In a Swedish screening study of five municipal sewage water treatment plants (STPs), the detection frequency of three FQs in digested dewatered sludge was 100% [13]. These substances have also been detected in hygienized sludge, i.e., dried digested sludge (above 90% dm), at Umeå STP [20]. In an extensive study by the U.S. Environmental Protection Agency, 84 sludge samples from 74 sewage treatment plants were investigated with the aim to determine four PAHs, 11 flame retardants, 72 pharmaceuticals, and 25 steroids and hormones [21]. Seen in relation to the total number of sludge samples that were analyzed, the detection frequency was as follows (in percentages): semi-volatile organic compounds and PAHs, over 50%; brominated flame retardants, 100%; 12 pharmaceutical compounds (including ciprofloxacin, diphenhydramine, and triclocarban), approximately

95%; and nine steroids, approximately 95%. In general, the levels of TCS and FQs in sludge are in the mg·kg⁻¹ dm range. The corresponding levels of PBDEs, PAHs, PCBs, and EE2 are in the ng kg⁻¹ dm range [12,13,14,15,21].

Information on the levels of synthetic organic substances in humus and mineral soil in sludge fertilized forest soils are scarce. In arable land, subjected to sludge fertilization, levels of PCBs and PBDEs were elevated, and bioaccumulation in earthworms was detected [22]. It has been shown that FQs may accumulate in soil if fertilization occurs regularly [12,23]. In general, several different biochemical and toxic effects, such as interference in reproduction, mutagenicity, and carcinogenicity, may occur in organisms exposed to TCS, PBDEs, PCBs, and PAHs. The main question regarding FQs and TCS is if their presence in the environment gives rise to resistant strains of bacteria, which in turn can produce changes in the ecosystem. Negative environmental effects of hormones have been reported, i.e., androgynous fish downstream of English STPs [24].

The aim of this study is to: (1) evaluate occurrence and distribution of selected synthetic organic substances in relevant matrices following application of dried and granulated municipal sewage sludge to boreal coniferous forest soils; and (2) perform an initial environmental risk assessment.

7.2 RESULTS AND DISCUSSION

7.2.1 LEVELS OF SYNTHETIC ORGANIC SUBSTANCES IN GRANULATED SLUDGE

The concentrations of the synthetic organic substances (see Table 1) in granulated sludge (in µg·kg⁻¹ dm) were as follows. FQ tot 6920; TCS 1170; EE2 1,4; PCB-7 41; PBDE tot 56; PAH-L 13; PAH-M 770; and PAH-H 750. FQs and TCS had the highest concentrations in the granulated sludge used for fertilizing L1. The levels of these four substances in the sludge are similar to those previously reported [12,13,15,20,21,25]. The EE2 concentration was approximately three orders of magnitude lower than those of FQs and TCS, but still in close correlation to previous findings in municipal sludge [26]. In addition, the levels of PBDEs (PBDE #209 not included in the PBDE tot of granulated sludge), PAHs, and PCBs were similar to those reported from the US Environmental Protection Agency [21]. For both PAHs and PCB-7, levels are below the recommended guide values, 3,000 and 400 $\mu g \cdot kg^{-1}$ dm, respectively, suggested being used within agriculture (according to a national agreement) [27].

7.2.2 LEVELS OF SYNTHETIC ORGANIC SUBSTANCES IN THE HUMUS LAYER AND MINERAL SOIL IN THE SMALL-SCALE EXPERIMENT (L1)

The extraction efficiency for the FQs from humus layers was inadequate, although this basic extraction method is suitable for sludge [13,20] and mineral soil. Although modifications of the presented method were made, including various strengths of EDTA and basic (or acidic) additives, added amounts of FQ substances were not recovered, at least not chemically intact, from the humus layer. Similar problems regarding extraction of antibiotics, including FQs, from solids and humic material have been addressed [29,30]. This is most likely due to a strong association of FQs to humus components. In house experiments, investigating the phase distribution of ciprofloxacin in water/humus matter systems, showed similar results. If this assumption is valid, the humus layer might act as a trap, in turn preventing the FQs from reaching the mineral soil and the soil water. The FQs were below LOQ in all of the mineral soil samples. The frequency of detection of EE2 was also low and it was only above LOQ in the dried granulated sludge.

TCS was detected at both sites fertilized by hand with dried granulated sludge, and the highest levels (humus) were found on the site with the highest dose (164 and 513 $\mu g \cdot kg^{-1}$ dm), shown in Figure 1. In the following two years, the concentrations of TCS have increased approximately three times, most likely due to further degradation of the granule itself. Following a total of four years, the TCS concentration in the fertilized humus layers have lowered, indicating that degradation or removal processes have occurred. However, the levels are still in close correlation to the findings of the first year of sampling. TCS was below LOQ in the mineral soil,

Table 1. The synthetic organic substances included in this study.

Substance	CAS[a]	Mw [a,b] (g mol-1)	Log P[a,c]	pK$_{a1}$[a,d]	pK$_{a2}$[a,e]
Antibacterial					
Triclosan (TCS)	3380-34-5	289.54	5.34	7,8	-
Antibiotics [f]					
Norfloxacin	70458-96-7	319.33	1.74	0.16	8.68
Ofloxacin	82419-36-1	361.37	1.86	5.19	7.37
Ciprofloxacin	85721-33-1	331.35	1.63	6.43	8.68
Hormone					
Ethinyl estradiol (EE2)	57-63-6	296.40	4.11	10.24	-
PBDEs [g]					
PBDE 47	5436-43-1	485.79	6.68	-	-
PBDE 99	60348-60-9	564,69	7.31	-	-
PBDE 183	207122-16-5	722.48	8.19	-	-
PBDE 209	1163-19-5	959. 17	9.45	-	-
PAHs [h]					
Acenaphtylene [i]	209-96-8	152.19	3.27	-	-
Acenaphthene [i]	83-32-9	154.21	3.73	-	-
Fluorene [j]	86-73-7	166.22	4.32	-	-
Phenanthrene [j]	1985-01-08	178.22	4.55	-	-
Anthracene [j]	120-12-7	178.22	4.55	-	-
Pyrene [j]	129-00-0	202.25	5.00	-	-
Fluoranthene [j]	206-44-0	202.25	5.00	-	-
Benz[a]anthracene [k]	56-55-3	228.29	5.73	-	-
Chrysene [k]	218-01-9	228.29	5.73	-	-
Benzo[b]fluoranthene [k]	205-99-2	252.31	6.19	-	-
Benzo[k]fluoranthene [k]	207-08-9	252.31	6.19	-	-
Benzo[a]pyrene [k]	50-32-8	252.31	6.19	-	-
Dibenzo[a,h]anthracene [k]	53-70-3	278.35	6.91	-	-
Indeno[1,2,3,cd]pyrene [k]	193-39-5	276.33	6.65	-	-
Benzo[ghi]perylene [k]	191-24-2	276.33	6.65	-	-
PCB-7 [L]					

Synthetic Organic Substances in Boreal Coniferous Forest Soils

Table 1. The synthetic organic substances included in this study.[a]

PCB 28	7012-37-5	257.54	5.72	-	-
PCB 52	35693-99-3	291.99	5.83	-	-
PCB 101	37680-73-2	326.43	6.44	-	-
PCB 118	31508-00-6	326.43	6.77	-	-
PCB 138	35065-28-2	360.88	6.98	-	-
PCB 153	35065-27-1	360.88	7.04	-	-
PCB 180	35065-29-3	395.32	7.51	-	-

[a] Reference [28]; [b] Molecular weight; [c] Octanol/water partition coefficient; [d] Acid dissociation constant of first proton; [e] Acid dissociation constant of second proton; [f] Fluoroquinolones (FQs); [g] Polybrominated biphenyl ethers; [h] Polyaromatic hydrocarbons; [i] Low molecular weight; [j] medium molecular weight; [k] high molecular weight, carcinogenic; [l] Indicator PCBs. The seven congeners usually analyzed, constitute 10%–30% of total PCB-content (in total 209 congeners).

suggesting that this substance's ability to reach underlying soil is limited, at least during the first two years following fertilizing.

The presence of PBDEs, PCBs, and PAHs in forest soils, in comparison to FQs, TCS, and EE2, is different since atmospheric deposition may give rise to background levels. In order to assess an increase of the concentration of such substances in fertilized soils, their levels should exceed those found in the control sites. The PBDEs and PCB-7 have, in parity with TCS, elevated levels in humus of the fertilized sites and it is clearly demonstrated in year two (Figure 2, Figure 3, respectively). The concentrations of PBC-7 of the fourth year sampling of the humus layer are, as in the case of TCS, lower than the results of the second year of sampling. In the fertilized mineral soil samples, the levels of PBDEs are in close correlation to those found in the control sample. The corresponding levels of PCB-7 were below 0.4 μg·kg^{-1}. Of the three groups of PAHs, none showed significant change following fertilizing, regardless of soil matrix, and instead, their levels were in close connection to those found in the controls, see Figure 4.

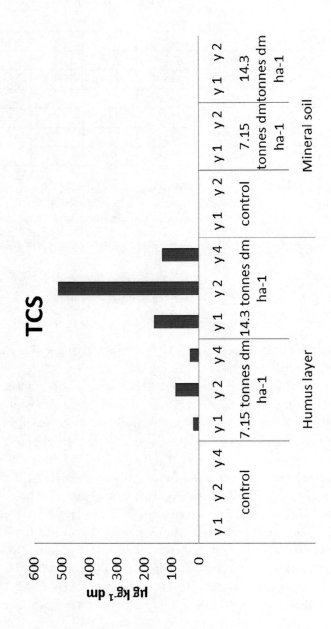

Figure 1. TCS levels in the small-scale experimental sites fertilized by hand (L1).

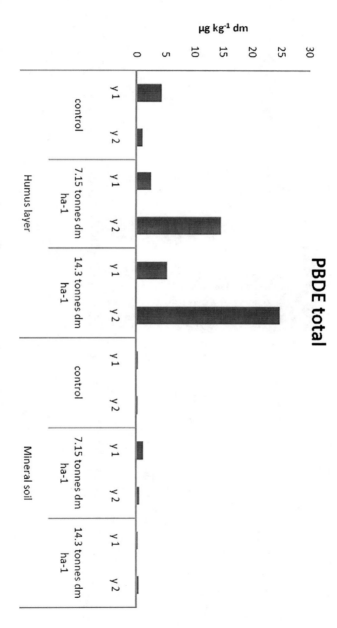

Figure 2. PBDE total levels in the small-scale experimental sites fertilized by hand (L1).

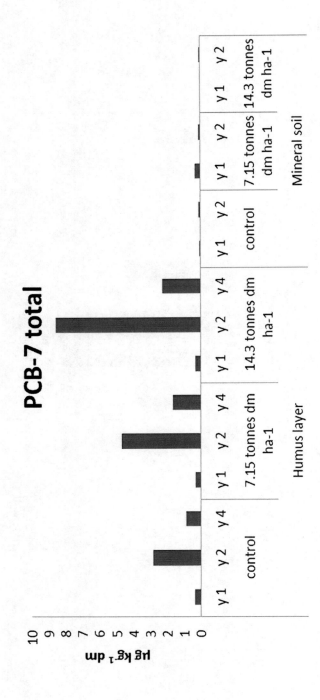

Figure 3. PCB-7 total levels in the small-scale experimental sites fertilized by hand (L1).

Synthetic Organic Substances in Boreal Coniferous Forest Soils 131

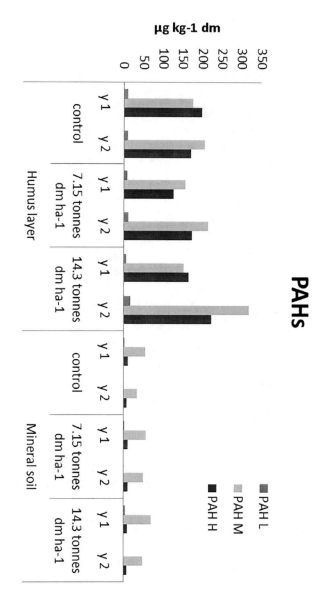

Figure 4. PAH levels in the small-scale experimental sites fertilized by hand (L1).

7.2.3 LEVELS OF SYNTHETIC ORGANIC SUBSTANCES IN THE HUMUS LAYER AND MINERAL SOIL IN THE FULL-SCALE SITES (L3-L4)

In Table 2, the results from the L3 and L4 sites are shown. FQs and EE2 were not found above their LOQ in any of the samples collected from full-scale sites (L3 and L4), including soil water (only FQs). TCS was present in all humus samples exposed to dried granulated sludge, and the highest concentration was seen in L4. Lower levels of TCS were seen in L3, but a positive result was obtained from a site not being fertilized. This could be explained by atmospheric deposition of TCS, however it is questionable if TCS is distributed in the environment via air. TCS was below LOQ in all but one of the mineral soil samples. PBDEs were present in all but one of the samples collected from the full-scale experimental locations (mineral soil sample from L4) and slightly elevated levels could be seen in the humus layer samples exposed to sludge. By comparison to the small-scale (L1) location, the levels of PBDEs in the full-scale (L3-L4) locations were slightly lower. In location L4 the concentrations of PCB-7, PAH-M, and PAH-H were approximately two to five times higher in the humus layer exposed to sludge compared to the control equivalents. Elevated levels of these three groups of synthetic organic substances could not be seen in the mineral soil. In addition, the results from the air sampling campaign (in L2) did not reveal any significant increases of the levels of the synthetic organic substances following full scale fertilizing by machine, regardless of the sample type (Table 3).

7.2.4 AVERAGE CONCENTRATIONS OF THE SYNTHETIC ORGANIC SUBSTANCES IN THE HUMUS LAYER

The average concentrations and standard deviations of the synthetic organic substances were calculated, see Figure 5. The results were based on: humus samples, control and fertilized sites (n = 5 of each), and all of the locations (L1-L4) included in this study (the results of L1 were based

Table 2. Levels (μg kg^{-1} dm) of the synthetic organic substances in samples from the full-scale sites.

Substances	L3[a]								L4[b]			
	H[c]	H	H[d]	H[d]	M[e]	M	M[d]	M[d]	H	H[d]	M	M[d]
FQ tot	*f*	*f*	*f*	*f*	*g*	*g*	*g*	*g*	*f*	*f*	*g*	*g*
TCS	9	*g*	31	18	*g*	*g*	17	*g*	g/gh	778/69[h]	*g*	*g*
EE2	*g*	*g*	*g*	*g*	*g*	*g*	*g*	*g*	*g*	*g*	*g*	*g*
PBDE tot	0.16	2.1	0.67	3.3	0.03	0.20	0.35	*g*	0.001	0.07	0.08	0.06
PCB-7	2.77	3.36	2.69	4.53	0.09	0.10	0.69	0.08	5.79/2,9[h]	16.70/10,00[h]	0.10	0.14
PAH-L	11	8	1	*g*	*g*	*g*	*g*	*g*	12	21	2	1
PAH-M	215	205	4	289	2	4	9	1	196	525	16	15
PAH-H	212	216	2	293	2	1	5	1	234	1199	12	18

[a] Furuberget location, see Table 4; [b] Bäcksjön location, see Table 4; [c] Humic layer samples; [d] Fertilized site; [e] Mineral soil sample; [f] Not analyzed; [g] Below limits of quantification; [h] Results of the additional sampling conducted October 2011.

Table 3. The absolute amount (in ng) in the passive samplers (SPMD), levels in air (active air sampling) and in pine needles, in ng m^{3-1} and µg kg^{-1} wet weight, respectively.

Substances	SPMD [a]	SPMD [b]	Air [a]	Air [b]	Pine needles [a]	Pine needles [b]
FQ tot	c	c	c	c	d	d
TCS	c	c	c	c	d	d
EE2	c	c	c	c	d	d
PBDE tot	4.3	8.2	0.008	0.005	2.42	0.99
PCB-7	3.7	2.9	0.03	0.02	0.17	0.12
PAH-L	0	1	0.5	0.4	0.4	0.5
PAH-M	92	64	1.4	0.8	13	16
PAH-H	29	59	0.2	0.1	7	4

[a] Before fertilizing; [b] Following fertilizing; [c] Not analyzed; [d] Below limits of quantification.

Synthetic Organic Substances in Boreal Coniferous Forest Soils   135

on concentrations obtained from the year two sampling campaign). The average concentrations were, in all but one case, namely PAH-L, elevated in sites fertilized with dried and granulated sludge and the most significant increase were seen in TCS and PBDEs concentrations. The immense standard deviations are most likely attributed to the degradation of the sludge granule, and/or the synthetic substances, as the time between fertilizing and sampling range two to four years. In addition, the composition of the sludge collected from the Himmerfjärden STP during the different times (years) may vary. The Swedish guide value of PCB-7, concerning sensitive land use (8 $\mu g \cdot kg^{-1}$ dm), is also shown in Figure 5 and this value is equivalent to twice the background levels of PCB-7 obtained in this study [31]. In two of the fertilized locations, L1 year two and L4, obtained concentrations exceeding this value, suggesting that this soil is contaminated and restrictions, in terms of use and/or activity, should be practiced. Additional guide values, relevant to this study, are those of PAH-L (3 $mg \cdot kg^{-1}$ dm), PAH-M (3 $mg \cdot kg^{-1}$ dm), and PAH-H (1 $mg \cdot kg^{-1}$ dm), and in one of the sampling sites (location L4), the PAH-H concentration exceeded its guide value.

7.2.5 ENVIRONMENTAL RISK ASSESSMENT

If humus, in fact, acts as a trap for the FQs, in combination with concentrations in mineral soil and ground water below the LOQ, the negative environmental impact of this group of substances may be negligible. However, it is unknown if the humus layer also disables the antibacterial properties of the FQs, in turn eliminating the risk of developing resistance strains of soil bacteria. Relevant studies investigating the effects of FQs on soil-living organisms are scarce, but several studies with concerns regarding the aquatic environment have been reported. The effects of ciprofloxacin on the bacterial community structure, with a focus on the mineralization of pyrene in marine sediment, yielded a calculated EC_{50} of 0.4 $mg \cdot kg^{-1}$ dm [32]. In other studies, plant uptake of FQs from the aquatic media were investigated and the conclusions were that this mode of action could not be excluded [33,34]. EE2 was only successfully quantified in the granulated

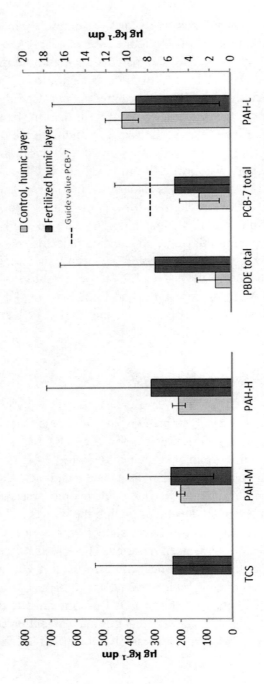

Figure 5. Average concentrations and standard deviations of the synthetic organic substances in control and fertilized humus samples.

sludge, and levels needed to develop an ecotoxicological response are close to the LOQs reported in this study [35]. Although many studies of the negative environmental impact of hormones exist, focus is attributed to organisms present in the aquatic media. Information regarding degradation of EE2 in soil matrices and its uptake in plants and bioaccumulation is missing. The risk associated with PAHs and fertilizing forest soils with granulated sludge, is most likely low since elevated levels of the PAHs, in contrast to the levels already present in the soil due to atmospheric deposition, could not be seen.

The environmental risk, associated with fertilization of granulated sludge, was considered possible for TSC, PCB-7, and PBDEs, since the levels of these three groups of substances were significantly elevated in humus exposed to sludge (in comparison the equivalent controls). In addition, the concentrations of the toxic, bioaccumulative, and persistent PCB-7 exceeded the established guide value for contaminated soil in two cases (in humus of location L2 year two and L4). Although guide values of PBDEs in soil matrices is missing, and relevant ecotoxicological, bioaccumulation, and degradation studies of TCS in the soil environment are missing, the potential risk of environmental effects should be considered. In this study the fertilized sites have been exposed to granulated sludge for a maximum of five years. The long-term effect is approximately 15 years when using granulated sludge and, seen in that perspective, it is not unlikely that the levels of TCS, PCB-7, and PBDEs (seen in the L1 location) will persist due to further degradation of the granule itself.

In an EU report from 2001, regarding organic contaminants in sludge used in agriculture, the available information at that time was compiled [36]. In the report, it is considered that the sludge is an important resource in order to return nutrients to agriculture. However, since sludge is the alternative holding the highest amounts of organic pollutants, the use of sludge cannot lead to adverse effects. The uptake of synthetic organic substances (e.g., PAHs, PCBs, PBDEs) in plants from soil is considered to be low but the uncertainty about other substances is high. The input of organic substances must be lower than their output (degradation), thus, it is possible to regulate the dose of various types of sludge. PCBs were identified as one of the candidates that should be monitored closely due to their vast degradation times. The level of knowledge regarding pharmaceuticals (including hormones) was assessed as very low and a proposal was that a permanent (at least 30-year-old) study about the persistence of organic compounds should be implemented prior to final recommendations. Not mentioned in the report are that synergistic effects between various organic substances (including metals) may occur and this phenomenon is not identified in toxicity tests targeting individual substances.

7.3 EXPERIMENTAL

7.3.1 SUBSTANCES OF INTEREST

The substances prioritized in this study can be seen in Table 1. The substances represent traditional persistent organic pollutants (PCBs, PAHs), as well as emerging environmental hazards (PBDEs and pharmaceuticals). The following substances are grouped and their individual concentrations are summed: FQs, PAH-L (low molecular weight), PAH-M (medium molecular weight), PAH-H (high molecular weight), PCB-7 (indicator PCBs), and PBDEs, see Table 1.

7.3.2 SLUDGE FERTILIZED FOREST SOILS AND SAMPLING

Samples were collected at four different sites in northern Sweden, see Table 4 for WGS84 coordinates and location characteristics. One of the sites (L1) was a small-scale experiment specifically prepared for this study. L1 was divided into three small test sites (10×10 m) and each site was fertilized by hand to ensure maximum control of the given dose. The dose of the dried granulated sludge (Himmerfjärden STP, Stockholm, Sweden) added to the three sites corresponded to (in tons dm·ha^{-1}): 0 (control); 7.15; and 14.3. The three test sites were fertilized in September 2007 and the sampling was conducted in November of 2008 and 2009. Additional sampling of the humus layers, focusing on TCS and PCB-7, was performed during October 2011. From the top humus layers, five samples were collected and

Synthetic Organic Substances in Boreal Coniferous Forest Soils

Table 4. Information regarding the sampling locations.

Location	Lat. [a]	Long. [a]	Altitude (m)	Vegetation	Trees	Age [b] (years)
(L1) Snårberget	66,269039°	23,099030°	146	Bilberry	Pines	48
(L2) Räktjärvberget	66,153964°	22,898668°	130	Bilb./Ling.	Pines	46
(L3) Furuberget	66,545295°	22,571772°	95	Lingonberry	Pines	60
(L4) Bäcksjön	63,944599°	20,410820°	70	Bilberry	Pines	70

[a] According to WGS84; [b] Age is determined by counting the annual rings on an increment core, taken at 1,3 m height, with an addition of 10 years for the trees in order to reach 1,3 m height.

pooled during 2008, 2009, and 2011. From the mineral soil, five samples were collected and pooled during 2008 and 2009, at a depth of approximately 30 cm. This strategy was performed at all of the three sampling sites at L1. In addition, the granulated sludge used to fertilize L1 was collected 2008.

Active air sampling was conducted in L2 24 h before and 24 h during/following fertilizing by machine (in August 2008). Pine needles were collected two weeks before and two weeks following fertilizing. Passive air sampling, by means of semi-permeable membrane devices (SPMDs), was conducted three weeks before and three weeks following fertilizing.

Humus and mineral soil (down to 30 cm depth) were gathered by pooling five subsamples (5 × 100 g) located in the middle, and in the four corners of a 2,000 m^2 square, within L3. The sampling was made two years following fertilizing by machine (August 2006). The doses of the granulated sludge were 0 (control) and 19.8 tons dm·ha^{-1}.

Humus and mineral soil (at 30 cm depth) were gathered by pooling five subsamples (5 × 100 g) located in the middle and in the four corners of a 2,000 m^2 square, within L4. The sampling was made five years following fertilizing by hand (Spring 2003). Five humus layer samples were also collected and pooled in October 2011. In addition, a soil water sample was also collected by the use of lysimeters from the control site and from the fertilized site. The doses of granulated sludge were 0 (control) and 13.6 tons dm·ha^{-1}.

7.3.3 CHEMICAL ANALYSIS

Prior to extraction of humus (cut to a maximum of 2 cm) and mineral soil samples, stones, and sludge granules were removed if needed.

Fluoroquinolones (FQs): Isotopic labeled ciprofloxacin ($^{13}C_3$, ^{15}N) was added to samples (2 g portions) of whole pine needles, mineral soil, and granulated sludge, to an amount of 1,000 ng per sample. The samples underwent liquid/solid extraction during 1 h with the use of 50% 0.1 M EDTA and 50% triethylamine (5%) in methanol/water (1:1). The sample extracts were spun at 5,000 rpm during 20 min and 1 mL of the supernatant was subjected to instrumental analysis. To the filtrated (0.45 μm) and acidified (pH 3) soil water samples (200 mL), 500 ng of the isotopically labeled ciprofloxacin was added and the samples were subjected to solid phase extraction (SPE, Isolute ENV+). The eluent was 5% triethylamine in methanol and the sample extracts were evaporated and re-dissolved in acidified (pH 3) acetonitrile/water (5/95). The sample extracts of the solid and aqueous phases were injected and analyzed by liquid chromatography electrospray tandem mass spectrometry (LC-ESI-MS/MS, LCQ DUO, Thermo Finnigan) [13]. The limits of quantification (LOQ) of solid and aqueous samples were 10 $\mu g \cdot kg^{-1}$ dm and 5 $ng \cdot L^{-1}$, respectively.

Triclosan (TCS): Isotopic labeled triclosan (^{13}C) were added (100 ng) to 2 g portions of of pine needles, humus layer, mineral soil, and granulated sludge samples. The samples were extracted 30 min in 15 mL methanol assisted by ultrasonication. The sample extracts were evaporated to approximately 1 mL and spun at 10,000 rpm during 20 min. The sample extracts were injected and analyzed by LC-atmospheric pressure photo ionization-MS/MS (TSQ Ultra, Thermo Electron) and the LOQ was 2.5 $\mu g \cdot kg^{-1}$ dm.

Ethinyl estradiol (EE2): Isotopic labeled EE2 ($^{13}C_2$) were added (500 ng) to 2 g portions of of pine needles, humus layer, mineral soil, and granulated sludge samples. The samples were extracted 30 min in 50 mL hexane/acetone (1:1) assisted by ultrasonication. The sample extracts were evaporated to dryness following extraction and re-dissolved in 10 mL acidified. The extracts were passed through SPE (Oasis HLB) before derivatization (silylation). The final sample extracts were injected and analyzed by gas chromatography high-resolution mass spectrometry (GC-HRMS, Micromass Ultima Autospec Ultra, (Waters Corp.). The LOQ was 1.2 $\mu g \cdot kg^{-1}$ dm.

Polyaromatic hydrocarbons (PAHs), polychlorinated biphenyls (PCBs), and polybrominated biphenyl ethers (PBDEs): Isotopic labeled analog representatives within each group were added to filter and polyurethane foam (from the active air sampling device), SPMD extracts and to 10 g portions of pine needles, humus layer, mineral soil, and granulated sludge. The samples were soxhlet extracted during 24 h in toluene, with the exception of pine needles that were extracted with dichloromethane and assisted by ultrasonication. The extracts were passed through multisilica (sulphuric acid, neutral, sodium hydroxide) and fluorisil, and the samples were injected and analyzed by GC-HRMS (Micromass Ultima Autospec Ultra, Waters Corporation). The LOQs (in ng kg^{-1}) were: PCB 0.1; PBDE 1; and PAH-L 10; PAH-M 20; and PAH-H 40.

7.4 CONCLUSIONS

This study has shown that following fertilizing with dried granulated sludge, elevated levels of synthetic organic substances (mainly TCS, PBDEs, and PCB-7) occur in the humus layer. The correlation between the dose of the sludge used and the obtained concentrations were stronger in the small-scale experimental sites (100 m^2), fertilized by hand, in comparison to the full-scale sites (several hectares), fertilized by machine. In some locations, the levels of PCB-7 slightly exceeded the guide value set by the Swedish Environmental Protection Agency (8 µg·kg^{-1}). The guide value was only exceeded for sludge application rates above the proposed application rate (10 tons·ha^{-1}) during forest fertilization in practical scale. Elevated levels of the synthetic organic substances were not found in the mineral soil, soil water, various types of samples related to air, or pine needles. Continuous monitoring of the synthetic organic substances should be conducted on the already fertilized locations. Even earlier experiments (if possible, more than 20 years ago) of forest soils, in which fertilization occurred with municipal sewage sludge, should be identified in order to improve the information regarding the levels, time trends, half-lives, and distribution of these substances in humus and mineral soils.

REFERENCES

1. Anon. On the Road to an Oil-Free Sweden; Report by the Government Persson Commission on Oil Independence. The Swedish Government: Stockholm, Sweden, 2006; p. 45.
2. Anon, A Forest Policy in Line with the Times; Government Bill: Stockholm, Sweden, 2007; Volume 108, p. 142.

Synthetic Organic Substances in Boreal Coniferous Forest Soils   143

3. Swedish Statistical Yearbook of Forestry; Swedish Forest Agency: Jönköping, Sweden, 2009.
4. Jacobson, S.; Kukkola, M. Forest Fuel Extraction in Thinning Gives Perceptible Growth Losses; Forestry Research Results. Skogforsk: Uppsala, Sweden, 1999; Volume 13, p. 4.
5. Recommendations for Extraction of Logging Residues and Wood Ash; Swedish Forest Agency: Jönköping, Sweden, 2008; Volume 2, p. 22.
6. Jacobson, S. Reversal of Ash Can Cause Growth Loss; Forestry Research Results. Skogforsk: Uppsala, Sweden, 1997; Volume 23, p. 4.
7. Action Plan for the Return of Phosphorus from Sewage; Swedish Environmental Protection Agency: Stockholm, Sweden, 2002; Volume 12, p. 204.
8. Sahlén, K. Growth effects after forest fertilization with dried granular / pelletized sewage sludge. In preparation.
9. Henry, C.L.; Cole, D.W.; Hinckley, T.M.; Harrison, R.B. The use of municipal and pulp and paper sludges to increase production in forestry. J. Sustain. For. 1994, 13, 41–55.
10. Börjesson, P. Life Cycle Assessment of Willow Production; Report. Faculty of Engineering, Lund University: Lund, Sweden, 2006; p. 60.
11. Sahlén, K. Sewage Sludge Fertilization of Conifer Forests in the Nordic Countries and North America; Nordic Council of Ministers: Copenhagen, Denmark, 2006; Volume 501, p. 67.
12. Golet, E.M.; Strehler, A.; Alder, A.C.; Giger, W. Determination of fluoroquinolone antibacterial agents in sewage sludge and sludge-treated soil using accelerated solvent extraction followed by solid-phase extraction. Anal. Chem. 2002, 74, 5455–5462.
13. Lindberg, R.H.; Wennberg, P.; Johansson, M.I.; Tysklind, M.; Andersson, B.A.V. Screening of human antibiotic substances and determination of weekly mass flows in five sewage treatment plants in Sweden. Environ. Sci. Technol. 2005, 39, 3421–3429.
14. Song, M.; Chu, S.; Letshcer, R.J.; Seth, R. Fate, partitioning, and mass loading of polybrominated diphenyl ethers (PBDEs) during the treatment processing of municipal waste. Environ. Sci. Technol. 2006, 40, 6241–6246.
15. Olofsson, U.; Lundstedt, S.; Haglund, P. Behavior and fate of anthropogenic substances at a Swedish sewage treatment plant. Water Sci. Technol. 2010, 62, 2880–2888.
16. Rapportering av uppdrag åt Naturvårdsverket, kontrakt 505 0302, Dnr 235–6086–03Me. Swedish Environmental Protection Agency: Stockholm, Sweden, 2004.
17. Swedish Chemical Agency. Polycykliska aromatiska kolväten (PAH). Available online: http://www2.kemi.se/templates/PRIOframes.aspx?id=4045&gotopage=4101/ (accessed on 19 June 2013).
18. Rapport Triclosan. Swedish Society for Nature Conservation: Stockholm, Sweden, 2007.
19. Swedish Association of the Pharmaceutical Industry. FASS. Available online: http://www.fass.se/LIF/miljo/miljoinfo.jsp/ (accessed on 19 June 2013).
20. Lindberg, R.H.; Olofsson, U.; Rendahl, P.; Johansson, M.; Tysklind, M.; Andersson, B. Behavior of fluoroquinolones and trimethoprim during mechanical,

144 Sewage and Landfill Leachate

chemical and active sludge treatment of sewage water and digestion of sludge. Environ. Sci. Technol. 2006, 40, 1042–1048.

21. TNNNS Sampling and Analysis Technical Report; EPA-822-R-08–016; US Environmental Protection Agency: Washington DC, USA, 2009.

22. Matscheko, N.; Tysklind, M.; de Wit, C.; Bergek, S.; Andersson, R.; Sellström, U. Application of sewage sludge to arable land—Soil concentrations of polybrominated diphenyl ethers and polychlorinated dibenzo-p-dioxins, dibenzofurans, and biphenyls, and their accumulation in earthworms. Environ. Toxicol. Chem. 2002, 21, 2515–2525.

23. Lindberg, R.H.; Björklund, K.; Rendahl, P.; Johansson, M.I.; Tysklind, M.; Andersson, B.A.V. Environmental risk assessment of antibiotics in the Swedish environment with emphasis on sewage treatment plants. Water Res. 2007, 41, 613–619.

24. Läkemedel och Miljö; Apoteket AB: Solna, Sweden, 2005.

25. Remberger, M.; Sternbeck, J.; Strömberg, K. Screening of Triclosan and Certain Brominated Phenolic Substances in Sweden. Swedish Environmental Research Institute: Stockholm, Sweden, 2002. Available online: http://www.ivl.se/downlo ad/18.7df4c4e812d2da6a416800071681/1300802485449/B1477.pdf (accessed on 19 June 2013).

26. Länsstyrelsen i Blekinges Län; County Council in Blekinge: Blekinge, Sweden, 2007.

27. Aktionsplans för återföring av fosfor ur avlopp. Swedish Environmental Protection Agency: Stockholm, Sweden, 2012; p. 5214. Available online: http://www. naturvardsverket.se/Documents/publikationer/620-5214-4.pdf (accessed on 19 June 2013).

28. SciFinder. Available online: https://scifinder.cas.org/scifinder/view/scifinder/scifinderExplore.jsf/ (accessed on 19 June 2013).

29. Yang, J.F.; Ying, G.G.; Zhao, J.L.; Tao, R.; Su, H.C.; Chen, F. Simultaneous determination of four classes of antibiotics in sediments of the Pearl Rivers using RRLC-MS/MS. Sci. Total Environ. 2010, 408, 3424–3432.

30. Tolls, J. Sorption of veterinary pharmaceuticals in soils: A review. Environ. Sci. Technol. 2001, 35, 3397–3406.

31. Swedish Environmental Protection Agency. Guideline values for contaminated land / Table of general guidelines. Available online: http://www.naturvardsverket. se/sv/Start/Verksamheter-med-miljopaverkan/Fororenade-omraden/Att-utreda-och-efterbehandla-fororenade-omraden/Riktvarden-for-fororenad-mark/Tabell-over-generella-riktvarden/ (accessed on 19 June 2013).

32. Näslund, N.; Hedman, J.E.; Agestrand, C. Effects of the antibiotic ciprofloxacin on the bacterial community structure and degradation of pyrene in marine sediment. Aquat. Toxicol. 2008, 90, 223–227.

33. Migliore, L.; Cozzolino, S.; Fiori, M. Phytotoxicity to and uptake of flumequine used in intensive aquaculture on the aquatic weed, Lythrum salicaria L. Chemosphere 2000, 40, 741–750.

34. Migliore, L.; Cozzolino, S.; Fiori, M. Phytotoxicity to and uptake of enrofloxacin in crop plants. Chemosphere 2003, 52, 233–1244.

35. Lai, K.M.; Scrimshaw, M.D.; Lester, J.N. The effects of natural and synthetic steroid estrogens in relation to their environmental occurrence. Crit. Rev. Toxicol. 2002, 32, 113–132.
36. Eur-Lex. 2013. Available online: http://eur-lex.europa.eu/LexUriServ/LexUriServ.do?uri=CELEX:31986L0278:EN:NOT/ (accessed on 19 June 2013).

PART IV

LEACHATE TREATMENTS

CHAPTER 8

Analysis of Electro-Oxidation Suitability for Landfill Leachate Treatment through an Experimental Study

ELENA CRISTINA RADA, IRINA AURA ISTRATE, MARCO RAGAZZI, GIANNI ANDREOTTOLA, AND VINCENZO TORRETTA

8.1 INTRODUCTION

The leachate generated by municipal solid waste (MSW) landfills contains a high level of organic and inorganic pollutants arising from the biological and physical-chemical processes within the controlled landfills. Leachate treatment is difficult, for a number of reasons [1]:

- High concentrations of organic and inorganic pollutants;
- Variability of the characteristics of the leachate over time, according to the dynamics of the biological degradation of the waste in the landfill (quality fluctuation) and the precipitation and other

---

© 2013 by the authors; licensee MDPI, Basel, Switzerland. Sustainability 2013, 5(9), 3960-3975; doi:10.3390/su5093960. Creative Commons Attribution license (http://creativecommons.org/licenses/by/3.0/). Used with the authors' permission.

hydrological balance factors, such as surface runoff, evapotranspiration, field capacity of the landfill, etc. (quantitative fluctuation).

The available treatment alternatives are classified as onsite treatments, which can be grouped as follows:

- Complete treatment i.e., reaching standards for discharging directly into surface waters;
- Pretreatment, i.e., reducing the quantity and/or polluting load of leachate, which is subsequently treated in an off-site plant and/or discharged into the sewage system.

In the last twenty years, several plants have been built to pretreat special refluents in medium-large treatment plants, managing significant quantities of leachate. Very often, these kinds of plants are also designed to treat other types of pollutants from industrial activities or from air washout [2,3]. Generally, after a chemical-physical pretreatment (sifting, chemical precipitation of metals, etc.) and, in some cases, a biological pretreatment (activated sludge, also combined with an ultra-filtration process—Membrane BioReactor, MBR), it is possible to feed the treated leachate into the wastewater line of a municipal plant.

In the case of mature leachates from MSW landfills, the greatest problem in achieving the regulatory levels set for discharging into the sewage system essentially involves the high concentration of ammonium associated with the low biodegradability of the chemical oxygen demand (COD).

In order to tackle this problem, much research has been done over the last ten years to identify innovative solutions for the treatment of liquid waste that contains high concentrations of refractory organic compounds and ammonium. Two types of treatment are of particular interest:

- Chemical electro-oxidation;
- Innovative biological removal.

In addition to the above mentioned solutions, it is also possible to use ammonia stripping plants, in both the gas and vapor phases, or ion

Analysis of Electro-Oxidation Suitability for Landfill Leachate Treatment   151

exchange resins, to remove the nitrates produced with electro-oxidation or with biological processes, or to use new solutions in chemical-physical treatments [4].

A conventional biological treatment could be adopted, however if biological nitrification/denitrification was applied, the high ammonium content of the leachate would require the addition of an external substrate and this would increase the cost of the treatment.

In the last ten years, for high ammonium concentrations a number of unconventional biological treatments have emerged (SHARON, Single Reactor High activity Ammonia Removal Over Nitrite; ANAMMOX, ANaerobic AMMonium OXidation, and CANON, Completely Autotrophic Nitrogen removal Over Nitrite) [5,6,7,8,9]. These solutions have already been adopted in full-scale plants [10]. In particular ANAMMOX (oxidation of ammonium under anaerobic conditions) has been studied since the mid 1990s at the University of Delft [11].

Considering the different alternatives proposed and described in the technical literature [12], taking into consideration the role of electrode materials [13] the priority of this experimental investigation was thus to study the treatability of the leachate using electro-oxidation, because of the simplicity of the plant and its management.

Electrochemical treatments [14,15,16] consist of applying a voltage that is constant over time (potentiostatic) to the refluent being treated, or alternatively, of passing a current that is constant over time (galvanostatic) through it. The electrolytic cell is the reactor in which this process takes place. As the continuous current flows, the negative electrode (cathode) gives electrons to chemical species, which then diminish; in contrast, the positive electrode (anode) receives electrons from chemical species, which become oxidized. A very important role is role of electrode materials is discussed together with that of other experimental parameters.

Both of these processes encourage the partial or total electrochemical oxidation reactions of the pollutants. Partial conversion is aimed at more easily biodegradable compounds, whilst total conversion is aimed at CO_2 and H_2O as final products. The choice of the electrode materials, voltages, currents and reaction times are decisive factors in the effectiveness of the treatment. Chemical electro-oxidation can be performed in two distinct ways: indirect or direct electro-oxidation [17,18,19].

An initial form of indirect electro-oxidation takes place in the presence of high concentrations of chlorides, Cl⁻; first of all, active chlorine is formed on the anode and, in rapid succession, hypochlorite HClO is produced, which has a strong oxidizing effect. This reaction is particularly suitable for saline refluent leachates. It removes many polluting molecules (for example, ammonium, which is partially transformed into nitrogen gas, N2); but there is a risk that organo-chloride sub-products may be formed [17,20,21,22,23].

A second type of indirect electro-oxidation leads to the production of hydrogen peroxide H_2O_2, which also has a strong oxidizing effect on recalcitrant organic substances. By adding bivalent iron, Fe_2^+ (electro-Fenton), the reaction may be further enhanced [24,25,26].

Another example of indirect electro-oxidation takes place in the presence of metal ions (Ag_2^+, Fe_3^+, Co_3^+, Ni_2^+) in a solution. These ions act as "mediators" and are oxidized onto the anode, passing from a low valence to a higher valence, thereby becoming much more reactive. They thus attack the organic compounds present in the solution and form hydroxyl radicals, which, in turn, are able to oxidize organic compounds. Many studies have demonstrated the efficiency of electrochemical technology in removing various kinds of pollutants [27,28,29,30,31]. Several experiments have been carried out to treat leachate by electro-Fenton. COD removal yields of even more than 85% were obtained. High yields were also observed for the removal of nitrogen [31,32,33,34].

The two main aims of this paper are to:

- Contribute to the knowledge of electrochemical treatments for the reduction of COD, biochemical oxygen demand (BOD_5), ammonium, and total suspended solids in leachate;
- Observe whether there was any hexavalent chromium in the liquid sample due to the different oxidative conditions of this treatment (considering that Cr^{III} could be transformed into Cr^{VI} by applying electro-oxidation). Differently from Cr^{III}, Cr^{VI} is a powerful carcinogenic [35], thus its formation must be avoided.

8.2 MATERIALS AND METHODS

8.2.1 DESCRIPTION OF THE LANDFILL

The landfill studied is located in Piedmont (northern Italy) and consists of three lots. Two lots are full and are in the post-cultivation phase; one lot is currently in use. The cultivation was carried out in lot 1 from 1988 until 1994, and in lot 2, from 1994 until 2000. The landfill receives approximately 25,000 tonnes of MSW per year.

The basin has bottom proofing, which consists of an HDPE sheet and an underlying monitoring layer. Proofing of the various parts of the landfill was carried out using HDPE sheets, with a high adherence, joined with a double layer. Currently the leachate is collected, pumped and stored in tanks and treated out from the landfill plant, in a specific wastewater and liquid waste treatment plant (conventional biological treatment with the third chemical finishing step).

Lots 1 and 2 are currently exhausted.

Lot 3 is wider than the first two, is physically separate from lots 1 and 2, and is basically a separate landfill in terms of storage capacity, leachate, rainwater, and road network. It was dug in three stages (called the 1st, 2nd and 3rd hydraulic sectors), from east to west, and the last stage was completed in 2006. From a management point of view, the most western point (which corresponds to the first stage and part of the second stage) is almost full, and the western part, which began operating following completion of the third stage, is currently being used. The overall area of lot 3 is approximately 30,600 m^2.

8.2.2 EXPERIMENTAL APPARATUS AND SAMPLE CHARACTERIZATION

The experimental apparatus used in the laboratory tests (Figure 1) consisted of:

- A rectangular reactor (20 × 10 × 10 cm), in transparent PVC, with a thickness of 10 mm;

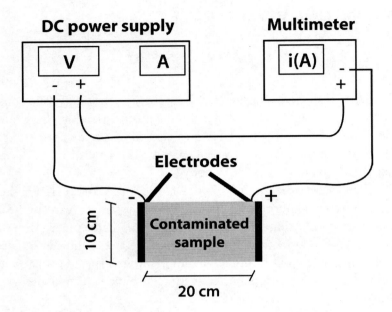

Figure 1. Experimental apparatus used in the tests.

- Two flat square electrodes (10 × 10 cm), made from stainless steel [36], which are perforated to allow the electro-osmotic mass to flow through;
- A DC power supply Mitek MICP 3005S (up to 60 V and up to 5 A).

When carrying out the laboratory experiments using this apparatus, the electrodes were inserted into the grooves on the walls of the empty reactor. The empty space between the electrodes was then filled with the contaminated sample. The leachate samples were analyzed in the laboratory. Regarding the intermediary and final sampling, because the sample was liquid, only one mixed sample that characterized the entire sample was taken and then preserved at 4 to 5 °C or taken directly to the laboratory for analysis.

The methods used for the analyses were MI-127 rev1 2007 (digestion/titration) for COD, MI-128 rev 2 2007 (respirometric method) for BOD, MI-122 rev 1 2007 (UV/visible spectro-photometry) for ammonium and hexavalent chromium, UNI EN ISO 14911:2001 for nitrates and nitrites, MI-155 rev 1 2007 (gravimetry) for TSS, and EPA 3052 2006 + EPA 6010 B 1996 for As, Be, Cd, total Cr, Ni, Pb, Cu, Se, and Zn.

Table 1 shows the characterization of leachate in the experimental tests. Table 1 highlights that both the Italian regulatory limits for metals and the local sewage company limits were respected, except for copper and zinc.

Table 1. Characterization of the leachate.

Parameter	Unit	Lot 1	Lot 2	Lot 3	Legal limits for discharge in sewage
COD	mg $O_2 \cdot L^{-1}$	3,580	3,360	4,314	12,000
BOD$_5$	mg $O_2 \cdot L^{-1}$	420	410	568	250
BOD$_5$/COD	-	0.117	0.122	0.132	-
Ammonia nitrogen	mg$\cdot L^{-1}$ N(NH$_4$)	1,728.60	2,181.1	2,296.3	3,500
Nitrite	mg$\cdot L^{-1}$ N-NO$_2$	< 0.01	< 0.01	< 0.01	0.6
Nitrate	mg$\cdot L^{-1}$ N-NO$_3$	0.10	2.64	0.16	30
Total suspended solids	mg$\cdot L^{-1}$	542	218	633	-
Chloride	mg$\cdot L^{-1}$	871	1,563	1,304	1,200
Arsenic	mg$\cdot L^{-1}$	0.01	0.1	0.09	0.5
Beryllium	mg$\cdot L^{-1}$	< 0.01	< 0.01	< 0.01	-
Cadmium	mg$\cdot L^{-1}$	< 0.002	0.003	0.003	0.02
Cobalt	mg$\cdot L^{-1}$	0.03	0.07	0.07	-
Nickel	mg$\cdot L^{-1}$	0.41	0.64	0.70	4
Lead	mg$\cdot L^{-1}$	0.017	0.026	0.142	0.3
Copper	mg$\cdot L^{-1}$	0.054	0.068	0.729	0.4
Selenium	mg$\cdot L^{-1}$	0.01	0.01	0.02	0.03
Zinc	mg$\cdot L^{-1}$	1.12	0.54	1.37	1.0
Total chromium	mg$\cdot L^{-1}$	0.499	0.856	1.372	4
Hexavalent chromium	mg$\cdot L^{-1}$	< 0.10	< 0.10	< 0.10	0.2

156 Sewage and Landfill Leachate

In any case, the main objective of this research did not include the treatment of metals, but rather the assessment of hexavalent chromium concentrations that can potentially change due to different oxidative conditions.

A comparison of data reported in Table 1 with literature data [37,38,39] demonstrates that the studied leachate is characteristic of an old landfill.

8.2.3 PARAMETERS

The parameters used in the treatments were adapted in accordance with the behavior of the matrix treated. All samples were characterized by high electrical conductivity values (between 14,500 $\mu S \cdot cm^{-1}$ and 25,500 $\mu S \cdot cm^{-1}$). It was therefore necessary to check the specific voltage values ($V \cdot cm^{-1}$).

In each test, a contaminated sample was introduced into the test cell, and a constant voltage, between 10 and 30 V, was applied for a set period of time (seven days for the first three samples from the three lots of the landfill, and one day for the fourth, made up of perforation water). At the end of each test, the sample was removed from the reactor and analyzed, to determine COD, BOD_5, ammonium, total suspended solid and hexavalent chromium values.

For the first three samples, the specific voltage applied varied between 1.5 $V \cdot cm^{-1}$ and 0.5 $V \cdot cm^{-1}$. During the first hour of treatment, the specific voltage was set at 1.5 $V \cdot cm^{-1}$, but the recorded current was too high (between 2 A and 3 A). In the next hour, the specific voltage was reduced to 1 $V \cdot cm^{-1}$. In the end, it was decided that the specific voltage used should be 0.5 $V \cdot cm^{-1}$.

The treatment period was seven days, with intermediate liquid samples and sludge taken after the first three days. In the literature, a maximum period of treatment is measured in hours (3–6 h) and not in days (as in this research), but the idea was to observe the behavior of the electrochemical treatment when no chemical substances were used.

We divided the treatment into two phases:

- Phase 1: lasting 3 days, application of a specific voltage of 0.5 $V \cdot cm^{-1}$;

- Phase 2: until the end of the test, with the application of a specific voltage of 1 V·cm^{-1}.

The idea of two phases with two different specific voltages was due to the fact that at the beginning of the test, the current value was too high and the power supply was not able to sustain a high voltage that would have corresponded to a specific voltage of 1 V·cm^{-1}. After three days, the current value dropped and the specific value of 1 V·cm^{-1} was established.

For the last sample, the treatment was applied only for one day, with a specific voltage of 1 V·cm^{-1}.

8.3 RESULTS AND DISCUSSION

The parameters monitored during the tests were: current, initial and final pH, initial and final electrical conductivity, and specific voltage. Table 2 shows the main characteristics of the tests carried out. In some cases, there was a peak in the current (which was occasionally very marked) in the initial moments of the test, followed by a gradual decrease. In all the tests, the current fell until it reached a level that remained constant for the remainder of the test, and then settled at very low values.

The current densities recorded were approximately 5.2 ÷ 8.8 mA·cm^{-2} (the surface of the electrochemical cell was about 200 cm^2 and the current had a variation between 1.04 A–1.76 A) at the beginning of the test, and fell significantly over time. These results differ slightly from the values reported in the literature [40] for electrochemical processes (1 mA·cm^{-2}).

8.3.1 LOT 1 TEST

This test was carried out for a treatment period of seven days, with a specific voltage that varied from 0.5 to 1 V·cm^{-1}.

During the initial hours of treatment, it was decided that a voltage of 1.5 V·cm^{-1} should be applied, however, given that, already after the first 30 minutes, the current value was greater than 2.8 A, the voltage was eventually reduced to 0.5 V·cm^{-1}. This voltage was kept constant for the first

Table 2. Summary of the lab tests carried out during the experiments.

Test	Applied voltage (V)	Specific voltage (V·cm⁻¹)	Duration (day)	Peak current (mA)
Lot 1	10 ÷ 20	0.5 ÷ 1	7	1,480
Lot 2	10 ÷ 20	0.5 ÷ 1	7	1,760
Lot 3	10 ÷ 20	0.5 ÷ 1	7	1,050

Table 3. Trend of the electrical current (test lot 1).

Time (h)	Current (A)	Specific voltage (V·cm⁻¹)
0.25	1.48	1
0.5	0.689	0.5
21	0.324	0.5
48	0.235	0.5
72	0.166	0.5
72.9	0.289	1
144	0.478	1
168	0.115	1

three days of treatment, after which, it was increased to 1 $V \cdot cm^{-1}$ in the second part of the treatment, because the current tended to fall to a constant value, which, in the case in question, did not exceed 170 mA. The values recorded during monitoring are shown in Table 3.

Taking into account the current values measured, the energy consumed was also assessed, and was found to be 740 $kWh \cdot m^{-3}$ (0.74 $kWh \cdot L^{-1}$).

By the end of the treatment, the pH increased slightly from its initial value of 7.5 to 7.7. The electrical conductivity fell from an initial value of 18.87 $mS \cdot cm^{-1}$, to 13.04 $mS \cdot cm^{-1}$, after seven days of treatment.

Table 4 shows the characterization of the sample, before, during and after the treatment.

Italian regulations for discharging into a sewage system establish require a maximum value for 271 COD of 500 $mg \cdot L^{-1}$ for COD. For the sample to conform with this, it is was necessary to apply the treatment for a time significantly longer than the seven days already tested, given that it

Analysis of Electro-Oxidation Suitability for Landfill Leachate Treatment 159

Table 4. Results of the Lot 1 sample treatment.

Parameter	Unit of measurement	Initial	Final	Treatment efficiency (%)
COD	mg O_2·L^{-1}	3,580	1,062	70.34
BOD	mg O_2·L^{-1}	420	133	68.33
Ammonia nitrogen	mg $N(NH_4)$·L^{-1}	1,728.6	1,066.2	38.32
Nitrite	mg $N-NO_2$·L^{-1}	< 0.01	< 0.01	-
Nitrate	mg $N-NO_3$·L^{-1}	0.1	611.79	-[a]
Total suspended solids	mg·L^{-1}	542	74	86.35
Hexavalent chromium	mg·L^{-1}	< 0.1	< 0.1	-

[a] The higher quantity of nitrate after the treatment was due to ammonia nitrogen transformation.

was demonstrated that the electrochemical treatment is was more effective with longer treatment times, but has not a linear kinetics [41].

After the first three days, sedimentation was observed, which had a volume of 0.3 L for lot 1.

From Table 4 it can be clearly observed that the efficiency of ammonium removal was low, the remaining ammonium concentration was about 1,000 mg·L^{-1}. The portion finally removed was 38.32%. The ammonium removed appeared to have been almost entirely transformed into nitrate.

An amount of 86.35% of total suspended solids had been removed from the liquid after the 7-day treatment.

8.3.2 LOT 2 TEST

This treatment lasted for seven days, with a specific voltage that varied between 0.5 and 1 V·cm^{-1}. The same treatment strategy described for the Lot 1 test was used. The trend of the current monitored during the test is shown in Table 5. The amount of energy consumed was 377 kWh·m^{-3} (0.38 kWh·L^{-1}).

Sewage and Landfill Leachate

Table 5. Trend of the electrical current (Lot 2 test).

Time (h)	Current (A)	Specific voltage (V·cm⁻¹)
0.17	1.76	1
0.5	0.763	0.5
21	0.27	0.5
48	0.097	0.5
72	0.069	0.5
72.25	0.658	1
144	0.083	1
168	0.072	1

Table 6. Results of the Lot 2 sample treatment.

Parameter	Unit	Initial	Final	Treatment efficiency (%)
COD	mg $O_2 \cdot L^{-1}$	3,360	1,176	65.00
BOD	mg $O_2 \cdot L^{-1}$	410	150	63.41
Ammonia nitrogen	mg $N(NH_4) \cdot L^{-1}$	2,181.1	1,851.9	15.09
Nitrite	mg $N\text{-}NO_2 \cdot L^{-1}$	< 0.01	< 0.01	-
Nitrate	mg $N\text{-}NO_3 \cdot L^{-1}$	2.64	209.64	-[a]
Total suspended solids	mg$\cdot L^{-1}$	218	246	-[b]
Hexavalent chromium	mg$\cdot L^{-1}$	< 0.1	< 0.1	-

[a] The higher quantity of nitrate after the treatment was due to ammonia nitrogen transformation.
[b] The increase in TSS is due to the electrode corrosion.

The pH value increased slightly by the end of the treatment: from its initial value of 7.5 to a final value of 7.7. The electrical conductivity fell from the initial value of 25.4 mS·cm⁻¹, to 17.21 mS·cm⁻¹ (after seven days of treatment).

The characterization of the sample, before, during and after the treatment is shown in Table 6.

A modest removal of COD and BOD was observed after the first three days of treatment, which continued to improve by the end of the treatment (from 25% to 65% for COD, and from 30% to 64% for BOD).

In view of the national maximum value (COD = 500 mg·L^{-1}) for discharging into the sewage system, the treatment was continued at least another 14 days (in addition to the seven already tested).

After the first three days, sedimentation could be seen, which, for lot 2, had a volume of 0.35 L.

For ammonium there was a modest final removal, of only 15.09%. In addition, in this case, ammonium seemed to be transformed into nitrate, for which an increase was recorded.

In this case, the total suspended solids did not register any removal, but an increase by nearly 13% after the 7-day treatment. This was due, in part, to the corrosion of the steel plates, with a subsequent release of suspended matter, and, in part, to electro-flocculation following the release of iron ions into the solution.

8.3.3 LOT 3 TEST

Similar methods were used for the test for Lot 3 (duration 7 days, voltage 0.5–1 V·cm^{-1}) (Table 7). The energy consumed amounted to 689 kWh·m^{-3} (0.689 kWh·L^{-1}).

A comparison of the current values for all three samples (Figure 2), reveals that even though the maximum value for the Lot 3 experiment was the lowest, the steady state value was the highest. In addition, for Lot 3 the last three measured values were very similar compared with the same values as in the Lot 1 and Lot 2 tests, where the variation is more obvious.

By the end of the treatment, pH increased slightly from its initial value of 7.5, to 8.1. In contrast, the electrical conductivity decreased from 26.7 mS·cm^{-1} to 19 mS·cm^{-1} (after seven days of treatment) (Table 8).

Here, too, in view of the limits regarding discharging into the sewage system, the treatment was continued for at least another 14 days (in addition to the seven already tested) to allow the sample to conform.

After the first three days, sedimentation was seen, with a volume of 0.35 L.

Table 7. Trend of the electrical current (Lot 3 test).

Time (h)	Current (A)	Specific voltage ($V \cdot cm^{-1}$)
0.09	1.05	1
0.5	0.337	0.5
21	0.361	0.5
48	0.194	0.5
72	0.094	0.5
72.5	0.563	1
144	0.629	1
168	0.607	1

Table 8. Results of the Lot 3 sample treatment.

Parameter	Unit of measurement	Initial	Final	Treatment efficiency (%)
COD	mg $O_2 \cdot L^{-1}$	4,314	1,506	65.09
BOD	mg $O_2 \cdot L^{-1}$	568	197	65.32
Ammonia nitrogen	mg $N(NH_4) \cdot L^{-1}$	2,296.3	902.1	60.72
Nitrite	mg $N-NO_2 \cdot L^{-1}$	< 0.01	< 0.01	-
Nitrate	mg $N-NO_3 \cdot L^{-1}$	0.16	993.29	
Total suspended solids	mg$\cdot L^{-1}$	633	383.6	39.40
Hexavalent chromium	mg$\cdot L^{-1}$	< 0.1	< 0.1	-

[a] The higher quantity of nitrate after the treatment was due to ammonia nitrogen transformation.

The final amount of ammonium removed was 60.72%. Here, too, it appeared to have transformed, in part, into nitrate, which registered an increase.

In this case, the total suspended solids registered a removal of almost 39.4% after the 7-day treatment. This increase could be attributed to corrosion of the steel electrodes, which may also occur in laboratory electro-oxidation tests carried out with steel plates. In real-scale plants, probes made of titanium activated with noble metals, are used, which are entirely

Analysis of Electro-Oxidation Suitability for Landfill Leachate Treatment

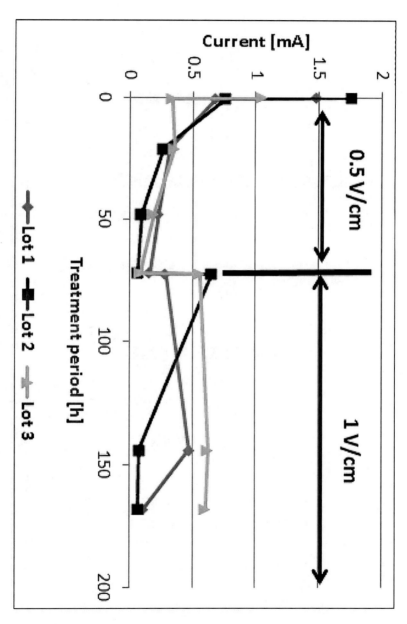

Figure 2. Comparison between current values for Lots 1, 2 and 3.

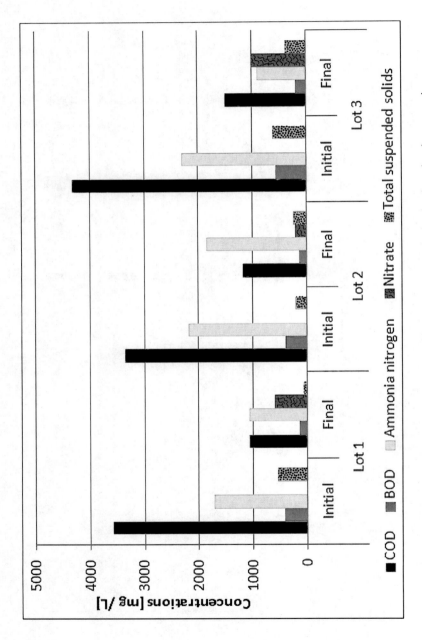

Figure 3. Comparison between the initial concentrations and the treatment results for the three leachate samples.

Analysis of Electro-Oxidation Suitability for Landfill Leachate Treatment 165

free of corrosion. The increase in the suspended solids was also due also to electro-flocculation, caused after iron ions were released into the solution.

8.3.4 RESULTS OF THE LEACHATE SAMPLES

Figure 3 compares the initial and final characterizations, for the three treated leachate samples.

Figure 4 and Figure 5 compare the discharge limits set by Italian regulations (Leg. Dec. No. 152/2006) [42] with the samples before and after the 7-day treatment, respectively.

In none of the cases were the limits established for COD, ammonium and nitrate complied with. Furthermore, for the samples from Lots 2 and 3, the limits established for total suspended solids, total chromium, nickel and copper were not complied with, because of the mechanisms described above, which are only important on a laboratory scale, when stainless steel electrodes are used.

Sedimentation in all three treatment (Lot 1, Lot 2 and Lot 3) samples was observed which may be due to the electrode corrosion.

Another important aspect is that the rate of ammonia converted into nitrate resulted variable: 35% for Lot 1, 10% for Lot 2 and 43% for Lot 3.

Moreover, data of Cr^{VI} in Table 4, Table 6 and Table 8 demonstrate that there is no detectable conversion of Cr^{III} into Cr^{VI}, with a clear advantage compared to other phenomena involving these forms of Cr (i.e., combustion [43], soils with specific composition [44]).

Very high values of power consumption were found, about ten times higher than previously reported data [41]. This is a demonstration that electro-oxidation is very sensible to the materials adopted at the electrodes.

8.4 CONCLUSIONS

Problems related to landfills are of great importance, considering both the real environmental pressure, with particular focus on leachate production, and the social acceptation of the risks for human health [45]. Thus, focusing research on leachate management and treatment is extremely

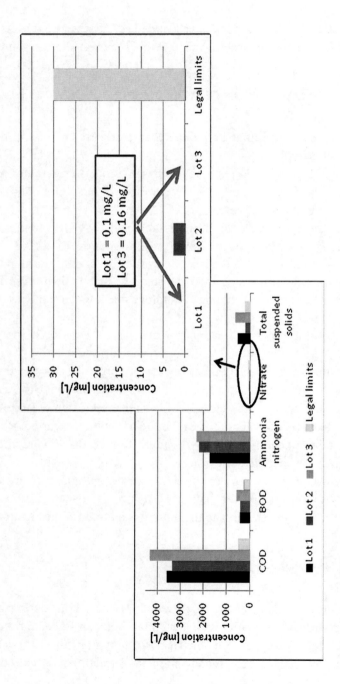

Figure 4. Initial concentrations of three leachate samples compared with regulation limits.

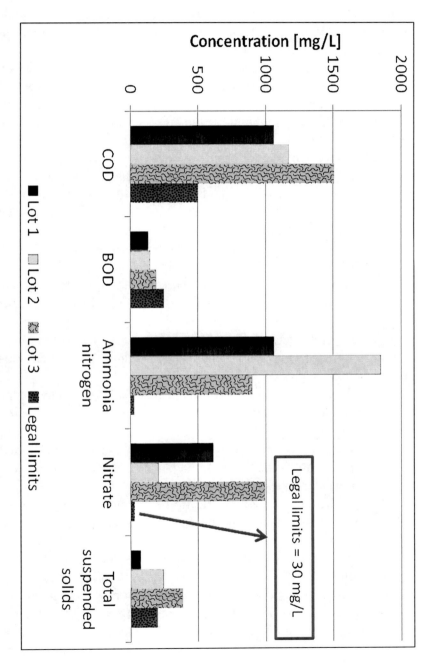

Figure 5. Final concentrations of the three leachate samples and comparison with the regulation limits.

important. We thus carried out tests on a laboratory scale, on a number of samples of landfill leachate, in order to verify the efficiency of electro-oxidation in terms of treatment.

The yields obtained are satisfactory, particularly considering the simplicity of the technology. As with all processes for treating wastewater, the applicability of this technique to a specific industrial effluent needs to be supported by feasibility studies that estimate its effectiveness and optimize the design parameters.

Electrochemical oxidation easily removed 64%–70% of COD, and 15%–61% of ammonium. The intervention was helped by the electrical conductivity, which had a fairly high value. In order to achieve the discharging limits for COD, and particularly those for nitrogen, electro-oxidation alone is not sufficient, and other treatments are needed.

Energy consumption for electro-oxidation treatment varied between 0.377 and 0.740 kWh·L^{-1}, depending on the treated sample. These values are only indicative, as they relate to laboratory tests, in which the optimal choice of electrode materials and of reaction times has not yet been obtained. In the literature, results related to energy consumption vary between 0.012 kWh·L^{-1} and 0.039 kWh·L^{-1} [46]. The selection of the anodic material based on cost limitation demonstrated that the energy consumption can increase significantly.

During the first few days of treatment, sedimentation of the samples was observed, which was usually equal to a third of the initial sample volume.

One problem was the presence of a higher value for some metals at the end of the treatment than at the beginning, which may have been due the corrosion of the electrodes.

In light of the tests carried out, electro-oxidation technology could be applied to leachates, to reduce the concentration of refractory organic matter and ammonium.

However, using this technology alone it appears that it is not possible to achieve the limits for discharging into sewerage, except where local limits for carbonaceous substances and nitrogen are decidedly more permissive.

Although a high energy consumption and a potential chlorinated organic formation may limit its application, electrochemical oxidation is a promising and powerful technology for the treatment of landfill leachate as was also stated by Deng and Englehardt in 2007 [26]. In general, it is reasonable to consider that electro-oxidation makes sense when renewable energy is used. This has recently been highlighted by several authors [47,48,49,50].

For the treatment of leachate, the most interesting solution might be to apply electro-oxidation in combination with hydrogen peroxide (electro-Fenton), when the main objective is to reduce the COD concentration under the discharging limits. However, when the intention is also to significantly reduce nitrogen, the easier alternative from a management point of view is to apply ammonia stripping in a washing tower, by pH conditioning. Alternatively, we suggest applying innovative, biological processes (ANAMMOX), which, despite limited application on a real scale internationally, are able to achieve around 80% nitrogen abatement.

Another limit to integrating electrochemical processes with ANAMMOX involves a negligible formation of nitrites.

In conclusion, the process sequence might be:

- Sequence 1: ammonia stripping + electro-oxidation (or electro-Fenton);
- Sequence 2: ANAMMOX, with an anaerobic membrane reactor, followed (if necessary) by electro-oxidation (or electro-Fenton). In this paper only initial and final values of the parameters characterizing the contaminant load are given; however, the process should be analyzed also on a kinetic basis and this is the natural development of the work. In fact a specific kinetic study is actually in project, after a feasibility study to verify in details the viability of the proposed hybrid treatment.

REFERENCES

1. Pi, K.W.; Li, Z.; Wan, D.J.; Gao, L.X. Pretreatment of municipal landfill leachate by a combined process. Process Saf. Environ. 2009, 87, 191–196.
2. Torretta, V. PAHs in wastewater: Removal efficiency in a conventional wastewater treatment plant and comparison with model predictions. Environ. Technol. 2012, 33, 851–855.

3. Torretta, V.; Katsoyiannis, A. Occurrence of polycyclic aromatic hydrocarbons in sludges from different stages of a wastewater treatment plant in Italy. Environ. Technol. 2013, 34, 937–943.
4. Luciano, A.; Viotti, P.; Mancini, G.; Torretta, V. An integrated wastewater treatment system using a BAS reactor with biomass attached to tubular supports. J. Environ. Manage. 2012, 113, 51–60.
5. Van Dongen, U.; Jetten, M.S.M.; van Loosdrecht, M.C.M. The SHARON ANAMMOX process for treatment of ammonium rich wastewater. Water Sci. Technol. 2001, 44, 153–160.
6. Notenboom, G.J.; Jacobs, J.C.; van Kempen, R.; van Loosdrecht, M.C.M. High Rate Treatment with SHARON Process of Waste Water from Solid Waste Digestion. In Proceedings of the IWA 3rd International Symposium Anaerobic Digestion of Solid Wastes, Munich, Germany, 11–13 September 2002.
7. Sliekers, A.O.; Third, K.A.; Abma, W.; Kuenen, J.G.; Jetten, M.S.M. CANON and ANAMMOX in a gas lift reactor. FEMS Microbiol. Lett. 2003, 218, 339–344.
8. Khin, T.; Annachhatre, A.P. Novel microbial nitrogen removal processes. Biotechnol. Adv. 2004, 22, 519–532.
9. Van Kempen, R.; ten Have, C.C.R.; Meijer, S.C.F.; Mulder, J.W.; Duin, J.O.J.; Uijterlinde, C.A.; van Loosdrecht, M.C.M. SHARON process evaluated for improved wastewater treatment plant nitrogen effluent quality. Water Sci. Technol. 2005, 52, 55–62.
10. Mulder, J.W.; Duin, J.O.J.; Goverde, J.; Poiesz, W.G.; van Veldhuizen, H.M.; van Kempen, R.; Roeleveld, P. Full-Scale Experience with the SHARON Process Through the Eyes of the Operators. In Proceedings of the WEFTEC 06, Dallas, TX, USA, 21–25 October 2006.
11. Mulder, A.; van de Graaf, A.A.; Robertson, L.A.; Kuenen, J.G. Anaerobic ammonium oxidation discovered in a denitrifying fluidized bed reactor. FEMS Microbiol. Ecol. 1995, 16, 177–184.
12. Moraes, B.P.; Bertazzoli, R. Electrodegradation of landfill leachate in a flow electrochemical reactor. Chemosphere 2005, 58, 41–46.
13. Martinez-Hiutle, C.A.; Ferro, S. Electrochemical oxidation of organic pollutants for the wastewater treatment: Direct and indirect processes. Chem. Soc. Rev. 2006, 35, 1324–1340.
14. Lei, Y.; Shen, Z.; Huang, R.; Wang, W. Treatment of landfill leachate by combined aged-refuse bioreactor and electro-oxidation. Water Res. 2007, 41, 2417–2426.
15. Istrate, I.A.; Grigoriu, M.; Badea, A.; Rada, E.C.; Ragazzi, M.; Andreottola, G. The Assessment of Chemical and Electrochemical Treatment for the Remediation of Diesel Contaminated Soils. In Proceedings of the Recent Advanges in Risk ManagementAssessment and Mitigation, Bucharest, Romania, 8–11 November 2010; pp. 144–149.
16. Oprea, I.; Badea, A.; Ziglio, G.; Ragazzi, M.; Andreottola, G.; Ferrarese, E.; Apostol, T. The remediation of contaminated sediments by chemical oxidation. Sci. Bull. 2009, 71, 131–142.
17. Bashir, M.J.K.; Isa, M.H.; Kutty, S.R.M.; Awang, Z.B.; Aziz, H.A.; Mohajeri, S.; Farooqi, I.H. Landfill leachate treatment by electrochemical oxidation. Waste Manage. 2009, 29, 2534–2541.

18. Bashir, M.J.K.; Aziz, H.A.; Aziz, S.Q.; Amr, S.S.A. An overview of electro-oxidation processes performance in stabilized landfill leachate treatment. Desalin. Water Treat. 2013, 51, 2170–2184.
19. Zhao, X.; Qu, J.; Liu, H.; Wang, C.; Xiao, S.; Liu, R.; Liu, P.; Lan, H.; Hu, C. Photoelectrochemical treatment of landfill leachate in a continuous flow reactor. Bioresource Technol. 2010, 101, 865–869.
20. Kurniawan, T.A.; Lo, W.; Chan, G.Y.S. Radicals-catalyzed oxidation reactions for degradation of recalcitrant compounds from landfill Leachate. Chem. Eng. J. 2006, 125, 35–57.
21. Deng, Y.; Englehardt, J.D. Electrochemical oxidation for landfill leachate treatment. Waste Manage. 2007, 27, 380–388.
22. Primo, O.; Rivero, M.J.; Urtiaga, A.M.; Ortiz, I. Nitrate removal from electro-oxidized landfill leachate by ion exchange. J. Hazard. Mater. 2009, 164, 389–393.
23. Uygur, A.; Kargi, F. Advanced treatment of landfill leachate by a new combination process in a full-scale plant. J. Hazard. Mater. 2009, 172, 408–415.
24. Do, J.S.; Yeh, W.C. Paired electro-oxidative degradation of phenols with in situ generated hydrogen peroxide and hypochlorite. J. Appl. Electrochem. 1996, 26, 673–678.
25. Robinson, H.D.; Knox, K.; Bone, B.D.; Picken, A. Leachate quality from landifilled MBT waste. Waste Manage. 2005, 25, 383–391.
26. Deng, Y.; Englehardt, J.D. Hydrogen peroxide-enhanced iron-mediated aeration for the treatment of mature landfill leachate. J. Hazard. Mater. 2008, 153, 293–299.
27. Li, X.Z.; Zhao, Q.L.; Hao, X.D. Ammonium removal from landfill leachate by chemical precipitation. Waste Manage. 1999, 19, 409–415.
28. Szpyrkowicz, L.; Kaul, S.N.; Neti, R.N.; Satyanarayan, S. Influence of anode material on electrochemical oxidation for the treatment of tannery wastewater. Water Res. 2005, 39, 1601–1613.
29. Cabeza, A.; Urtiaga, A.M.; Ortiz, I. Electrochemical treatment of landfill leachates using a boron-doped diamond anode. Ind. Eng. Chem. Res. 2007, 46, 1439–1446.
30. Mohan, N.; Balasubramanian, N.; Basha, C.A. Electrochemical oxidation of textile wastewater and its reuse. J. Hazard. Mater. 2007, 147, 644–651.
31. Lin, S.H.; Chang, C.C. Treatment of landfill leachate by combined electro-Fenton oxidation and sequencing batch reactor method. Water Res. 2000, 34, 4243–4249.
32. Zhang, H.; Zhang, D.; Zhou, J. Removal of COD from landfill leachate by electro-Fenton method. J. Hazard. Mater. 2006, 135, 106–111.
33. Atmaca, E. Treatment of landfill leachate by using electro-Fenton method. J. Hazard. Mater. 2009, 163, 109–114.
34. Mohajeri, S.; Aziz, H.A.; Isa, M.H.; Zahed, M.A.; Adlan, M.N. Statistical optimization of process parameters for landfill leachate treatment using electro-Fenton technique. J. Hazard. Mater. 2010, 176, 749–758.
35. U.S. Environmental Protection Agency, Toxicological Review of Hexavalent Chromium; U.S. Environmental Protection Agency: Washington, DC, USA, 1998.
36. Rusănescu, C.O.; Rusănescu, M. The stress-strain curves determined for microalloy steel with V determined on the torsion tests. Metalurgia (Bucharest) 2007, 59, 38–44.

37. Kulikowska, D.; Klimiuk, E. The effect of landfill age on municipal leachate composition. Bioresource Technol. 2008, 99, 5981–5985.
38. Zhang, Q.-Q.; Tian, B.-H.; Zhang, X.; Ghulam, A.; Fang, C.-R.; He, R. Investigation on characteristics of leachate and concentrated leachate in three landfill leachate treatment plants. Waste Manage, 2013. Available online: http://dx.doi.org/10.1016/j.wasman.2013.07.021 (accessed on 22 July 2013).
39. Frascari, D.; Bronzini, F.; Giordano, G.; Tedioli, G.; Nocentini, M. Long-term characterization, lagoon treatment and migration potential of landfill leachate: A case study in an active Italian landfill. Chemosphere 2004, 54, 335–343.
40. U.S. AEC, In-Situ Electrokinetic Remediation of Metal Contaminated Soils—Technology Status Report; Report No. SFIM-AEC-ET-CR-99022; Army Environmental Center: Fort Sam Houston, TX, USA, 2000.
41. Panizza, M.; Martinez-Huitle, C.A. Role of electrode materials for the anodic oxidation of a real landfill leachate—Comparison between Ti–Ru–Sn ternary oxide, PbO_2 and boron-doped diamond anode. Chemosphere 2013, 90, 1455–1460.
42. Italian Government. Italian Legislative Decree No. 152/2006—Environmental norms. Available online: http://www.camera.it/parlam/leggi/deleghe/06152dl.htm (accessed on 22 July 2013).
43. Stam, A.F.; Meij, R.; Winkel, H.T.; van Eijk, R.J.; Huggins, F.E.; Brem, G. Chromium speciation in coal and biomass co-combustion products. Environ. Sci. Technol. 2011, 45, 2450–2456.
44. Fendorf, S.E. Surface reactions of chromium in soils and water. Geoderma 1995, 67, 55–71.
45. Di Mauro, C.; Bouchon, S.; Torretta, V. Industrial risk in the Lombardy Region (Italy): What people perceive and what are the gaps to improve the risk communication and the partecipatory processes. Chem. Eng. Trans. 2012, 26, 297–302.
46. Ilhan, F.; Kurt, U.; Apaydin, O.; Gonullu, M.T. Treatment of leachate by electrocoagulation using aluminum and iron electrodes. J. Hazard. Mater. 2008, 154, 381–389.
47. Alvarez-Guerra, E.; Dominguez-Ramos, A.; Irabien, A. Design of the Photovoltaic Solar Electro-Oxidation (PSEO) process for wastewater treatment. Chem. Eng. Res. Des. 2011, 89, 2679–2685.
48. Alvarez-Guerra, E.; Dominguez-Ramos, A.; Irabien, A. Photovoltaic solar electro-oxidation (PSEO) process for wastewater treatment. Chem. Eng. J. 2011, 170, 7–13.
49. Dominguez-Ramos, A.; Aldaco, R.; Irabien, A. Photovoltaic solar electrochemical oxidation (PSEO) for treatment of lignosulfonate wastewater. J. Chem. Technol. Biot. 2010, 85, 821–830.
50. Valero, D.; Ortiz, J.M.; Expósito, E.; Montiel, V.; Aldaz, A. Electrochemical wastewater treatment directly powered by photovoltaic panels: Electro-oxidation of a dye-containing wastewater. Environ. Sci. Technol. 2010, 44, 5182–5188.

CHAPTER 9

Potential of Ceria-Based Catalysts for the Oxidation of Landfill Leachate by Heterogeneous Fenton Process

E. ANEGGI, V. CABBAI, A. TROVARELLI, AND D. GOI

9.1 INTRODUCTION

Landfill leachate is a liquid waste of primary environmental concern because of the quantity and quality of the harmful pollutants contained in it. There are a large number of various types of organic and inorganic substances, depending on the age and type of solid wastes located in the landfill. Leachate from sanitary landfills can be an important source of ground water contamination and for this reason it is collected from the bottom of the landfill to be treated; further, this highly contaminated liquid waste accumulates a great diversity of harmful pollutants. Some of them are particularly refractory and for this reason traditional wastewater treatment plants are not efficient in their abatement. Inorganic and organic content of leachate is characteristically related to environmental risk because of scarce biodegradation, severe bioaccumulation, and potential health damages [1, 2]. It is well known that conventional biological liquid waste treatments alone are unable to achieved complete removal of the leachate pollution over the life of the landfill. In truth conventional biological processes are time consuming and low-efficiency methods to treat directly leachate, consequently physicochemical processes are frequently utilized

---

© 2012 E. Aneggi et al.. International Journal of Photoenergy, *Volume 2012 (2012), Article ID 694721;* *http://dx.doi.org/10.1155/2012/694721. Creative Commons Attribution license (http://creativecommons.org/licenses/by/3.0/). Used with the authors' permission.*

174   Sewage and Landfill Leachate

to pretreat this liquid waste in order to reduce organic refractory before biological action in treatment plants units [3].

The most employed and studied methods in landfill leachate pretreatment are chemical or electrochemical coagulation [4], precipitation [5], and oxidation [6, 7]. Among these, a particular attention is given to oxidation techniques and especially to advanced oxidation processes (AOPs).

AOPs are methods able to convert nonbiodegradable organic pollutants into nontoxic biodegradable forms [8, 9], by the production of highly oxidizing hydroxyl radical species that promptly oxidize organic pollutants by a broad range of actions.

As a matter of fact oxidation by hydroxyl radicals species can be activated starting from H_2O_2 by intervention of transition metal salts (e.g., iron salts) [10], from ozone [11] or UV-light [12], leading to a more effective method to decompose certain refractory contaminants of leachate. In particular, Fenton oxidation is a well-known AOP used as pretreatment of leachate worldwide [10].

The Fenton's reagent works at mild temperature and pressure generating hydroxyl radicals following the generally accepted structure of reactions:

$$Fe^{2+} + H_2O_2 \rightarrow Fe^{3+} + OH^{\cdot} + OH^-$$
$$RH + OH^{\cdot} \rightarrow H_2O + R^{\cdot} \tag{1}$$
$$R^{\cdot} + Fe^{3+}\text{-oxidation} \rightarrow R^+ + Fe^{2+}$$

This reaction is followed by other very complex oxidation reactions in which a lot of radical forms are generated and take part in the overall Fenton oxidation. The H_2O_2 can act both as a scavenger or initiator, all organics in liquid waste can participate in radical generation [10] and the ferric iron catalyzes and decomposes H_2O_2 to additional radical forms contributing to the oxidation [13]. Moreover, the reaction of ferrous iron forms ferric hydroxo complexes which can contribute to the coagulation capacity of the Fenton reagent [14]. The reactions including hydrogen peroxide and ferric ions or other transition metal ions are also reported as Fenton-like reactions [15, 16]; moreover, some new wet peroxidations, in which various catalysts are added with hydrogen peroxide to remove organic compounds by low temperature reactions, are presented as heterogeneous Fenton-like systems [17–19].

The Fenton process is one of the most interesting AOPs when it is used to treat or pretreat heavily contaminated liquid wastes, and a lot of full-scale applications are installed over the world. The main advantage is to reach treatment of liquid wastes at mild conditions of temperature and pressure, but the most important drawback is the production of a sludge which needs to be treated as well. It is also a recognized concept that Fenton process, at reasonable reagents concentration, cannot lead to the complete mineralization of all organic compounds and often only partial oxidation occurs even in assisted oxidations [20].

Leachate treatment by classic Fenton process was often studied to assay potential increase of the biodegradability or reduction of toxicity or color removal [21–23]. Recently, photo-Fenton [24] and electro-Fenton [25] processes have been investigated for landfill leachate treatment and several studies have been dedicated to heterogeneous Fenton treatment of phenolic [19, 26, 27] or industrial wastewater [28–31]. Heterogeneous process could be a promising alternative due to the more important drawback of classic Fenton, the large amount of iron required for the reaction that dramatically exceeds the legally quantity permitted for effluent discharge ($<2\,mg/L$) and consequently requires a final waste management. At present, at the best of our knowledge, only one experience is reported about leachate treatment by heterogeneous catalytic Fenton-like systems [32], which can potentially be a promising way to activate radicals oxidizing species with minor sludge production. The experience reported in this paper tries to give a contribution to this theme.

The use of ceria-based materials in catalytic science is well established [33, 34]. Ceria is presently used in a large number of industrial processes and it accounts for a large part of the rare earth oxide market. Undoubtedly, its major commercial application is in the treatment of emissions from internal combustion engines where ceria-based materials have been used in the past 30 years [35]. Its more important action in three-way catalysts (TWCs) is to take up and release oxygen following variations in the stoichiometric composition of the feedstream; however, several other processes also benefit from the use of cerium and its derivates. Organo-soluble compounds of cerium are used as fuel additives for diesel engines and industrial boilers to reduce carbon deposits after combustion. Cerium oxide is used also as a catalytically active component to oxidize the liquid portion of particulate present in diesel engine exhaust. CeO_2 is also used as an additive or a promoter in commercial applications such as fluid catalytic cracking, ammoxidation and dehydrogenation processes [33–37]. Moreover, in the last years several ceria-based catalysts were investigated for CWAO (catalytic wet air oxidation) [38–40] and CWOP (catalytic wet peroxide oxidation) [41–44] techniques. The main purpose of this work is to investigate doped ceria materials in the treatment of landfill leachate by a heterogeneous Fenton process.

9.2 MATERIALS AND METHODS

9.2.1 CHARACTERIZATION AND SAMPLING OF LANDFILL LEACHATE

The leachate used in this study was drawn from an aged landfill near the city of Udine (northeast Italy); the landfill is not equipped with a recirculation system and it produces a stable leachate with high concentrations of COD and low BOD/COD ratio with a brown-green color. All the measures are carried out following the Standard Methods for the Examination of Water and Wastewater [45], COD was determined by colorimetric method utilizing a Hach Lange-DR 5000 spectrophotometer and quenching measurable residual H_2O_2 to prevent interference by addition of $MnO_{2(s)}$. Residual H_2O_2 was checked and determined to be zero by using test strips (Peroxide test sticks Quantofix, Sigma Aldrich). The TOC analyses were performed by a TOC-VCPN, Shimadzu analyzer.

9.2.2 PREPARATION AND CHARACTERIZATION OF THE CATALYSTS

The catalysts, ceria (CeO_2, CZ100), ceria-zirconia solid solutions (Ce0.44Zr0.56O2, CZ44) and Fe-doped materials ($Ce_{0.85}Fe_{0.15}O_{1.925}$, CF and $Ce_{0.45}Zr_{0.40}Fe_{0.15}O_{1.925}$, CZF) were prepared by coprecipitation starting from nitrates. Precipitates were dried at 393 K and calcined at 773 K for 2 h. Fe-doped materials were also calcined at higher temperature (1073 K) to investigate the behavior of iron-phase in the system. Textural characteristics of all samples were measured according to the BET method by nitrogen adsorption at 77 K, using a Tristar 3000 gas adsorption analyzer (Micromeritics).

Structural features of the catalysts were characterized by X-ray diffraction (XRD). XRD patterns were recorded on a Philips X'Pert diffractometer operated at 40 kV and 40 mA using nickel-filtered Cu-K$_\alpha$ radiation. Spectra were collected using a step size of 0.02° and a counting time of 40 s per angular abscissa in the range 20°–145°. The Philips X'Pert High-Score software was used for phase identification. The mean crystalline size was estimated from the full width at the half maximum (FWHM) of the X-ray diffraction peak using the Scherrer equation [46] with a correction for instrument line broadening. Rietveld refinement [47] of XRD pattern was performed by means of GSAS-EXPGUI program [48, 49]. The accuracy of these values was estimated by checking their agreement against the values of the lattice constant, assumed to comply with the Vegard's law [50].

In order to evaluate the oxygen/storage capacity (OSC) of samples TGA, experiments in Ar/H$_2$ (5%) flow (total flow 100 mL/min) were carried out. Each sample was treated in N$_2$ atmosphere for 1 h at 553 K. Then, it was heated at a constant rate (10 K/min) till 673 K and kept at this temperature for 15 minutes, to eliminate the absorbed water. Finally, Ar/H$_2$ mixture was introduced while keeping the temperature at 673 K for 30 minutes. The observed weight loss is due to oxygen removal by H$_2$ to form water, and it can be associated to total oxygen storage capacity at that temperature [51, 52].

9.2.3 CATALYTIC ACTIVITY

9.2.3.1 HETEROGENEOUS FENTON

A pressure vessel (Parr Instruments) equipped with a glass batch reactor with continuous stirring (400 rpm) (Figure 1) was used to carry out

Fenton-like oxidative reactions. The experiments were conducted for 120 minutes at 343 K stirring 100 mL of leachate with 10 mg of catalysts and 5 mL of H_2O_2 (3%). At the end of the reaction (2 hours), samples were taken out and analyzed. Each experiment was repeated three times to obtain the reproducibility (error bars are included in figures).

9.3 RESULTS AND DISCUSSION

9.3.1 TEXTURAL AND STRUCTURAL CHARACTERIZATION

The leachate selected to test oxidative Fenton-like process was characterized by a small concentration of iron in the raw mixture, a high pH value, a slight high value of COD and TOC if compared to average values of other old landfill leachate [53]. The main properties are described in Table 1.

Textural and structural characterization of all catalysts is reported in Table 2. Materials have surface area in the range 55–135 m^2/g. Ceria-zirconia solid solutions (CZ44 and CZF) show higher surface area with respect to ceria-based samples (CZ100 and CF) due to the stabilization effect of zirconia.

The introduction of ZrO_2 significantly enhances textural properties, indeed, sintering in ceria-zirconia is less important in accordance with its better thermal resistance [54].

Doping ceria has a significant positive effect on the catalytic, oxygen storage/redox and thermal properties of catalysts. The introduction of Zr^{4+} induces a structural modification and this factor plays a key role in the redox behaviour of ceria-zirconia solid solutions. The substitution of Ce^{4+} with Zr^{4+} produces a contraction of the cell volume and induces stress in the structure and consequently structural defects that increase the oxygen mobility. It is important to point out that the oxygen mobility is increased if no modification in the structure of solid solution is observed. From these considerations, we noted that better performances are achieved for solid solutions with cubic symmetry and with a high level of Zr^{4+}. Alternatively, a higher amount of ZrO_2 decreases the number of redox sites and consequently the activity of the system. There is an inverse relationship between the two effects; in order to obtain an active system it is important to

Potential of Ceria-Based Catalysts for the Oxidation of Landfill Leachate

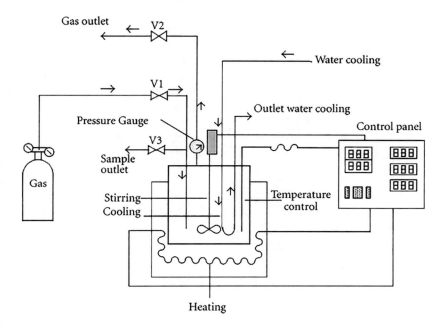

Figure 1. Schematic representation of the batch oxidation reactor used for tests.

balance the amount of structural defects and the amount of ceria. Literature data suggest that better results are obtained for compositions between CZ50 and CZ90 [55–58].

The structural features of all samples were analyzed by XRD.

In CeO_2-ZrO_2 system, several phases could be formed, depending on preparation conditions and concentration of single-oxide constituents [59]. In general, for a CeO_2 content <20 mol% a single-phase monoclinic cell is observed, while in CeO_2-rich compositions (CeO_2 > 70 mol%) solid solutions of cubic symmetry are formed. At intermediate levels, regions of tetragonal (t, t', and t'' phases) and cubic symmetry coexist in the phase diagram, their formation depending on the preparation method used. In our case, the Rietveld analysis of the diffraction profile of the materials

Table 1. Characterization of the landfill leachate used in this study.

Parameter	Unit of measurement	Values
pH	—	9
BOD_5	mg O_2/L	60
COD	mg O_2/L	2500
BOD5/COD	—	0.024
TN	mg N/L	1860
TOC	mg C/L	575
AOS	—	−2.52
ΔOD	mg O_2/L	0.38
Ammonia	mg NH_4^+/L	2150
Chloride	mg Cl^-/L	—
Color	PtCo unit	3600
Total iron	mg Fe/L	1.2
Nitrate	mg NO_3/L	—
Orthophosphate	mg PO_4^{3-}/L	60
Sulfate	mg SO_4^{2-}/L	—

has been carried out by opening the fitting to cubic, tetragonal and a mixture of the two.

As shown in Table 2, XRD measurements suggest that for binary ceria-zirconia samples with cerium content greater than 40 mol% the formation of a cubic fluorite lattice is favored, in accordance with the literature [60]. Thus, our ceria and ceria-zirconia solid solution crystallize in a cubic fluorite structure of *Fm3m* symmetry. In CZ44, no peak splitting that would indicate the presence of two phases could be detected, and therefore, the diffraction patterns demonstrate the formation of a single solid solution-like ceria-zirconia phase. This cannot exclude the presence of different arrangements of oxygen sublattice or the presence of a multiphase system at a nanoscale level, not detected by XRD. In fresh samples doped with

Table 2. Characteristics of catalysts used in this study and crystallographic parameters of modified ceria samples as obtained from Rietveld refinement and Vegard's law.

Sample	Composition	BET surface area (m^2/g)	Crystallite size (nm)[a]	Phase	Cell parameter $a = b = c$ (Å)	From Vegard's law
CZ100	CeO_2	53	7	Cubic	5.411 (1)	5.411
CZ44	$Ce_{0.44}Zr_{0.56}O_2$	90	4.7	Cubic	5.281 (1)	5.285
CF	$Ce_{0.85}Fe_{0.15}O_{1.925}$	77	7.5	Cubic	5.396 (1)	5.263
CF (1073 K)	$Ce_{0.85}Fe_{0.15}O_{1.925}$	22	31.1	Cubic	5.408 (1)	5.263
CZF	$Ce_{0.45}Zr_{0.40}Fe_{0.15}O_{1.925}$	132	3.5	Cubic	5.295 (1)	5.163
CZF (1073 K)	$Ce_{0.45}Zr_{0.40}Fe_{0.15}O_{1.925}$	22	8.9	Cubic	5.292 (1)	5.163

[a] Calculated with Scherrer formula from X-ray diffraction patterns.

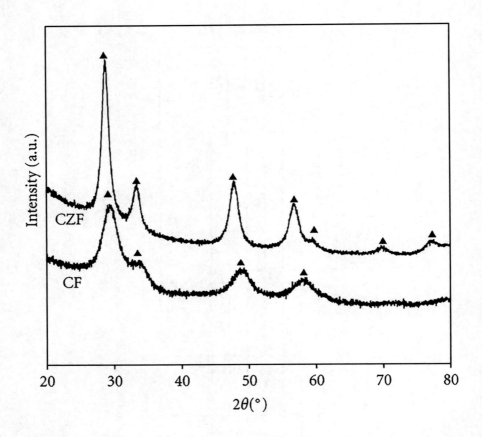

Figure 2. XRD profile for fresh (a) and aged (b) iron-doped samples (▲: CeO2 and CeZrO2; ■: Fe2O3).

Fe, XRD features allow to detect only the CeO_2 or $Ce_{0.44}Zr_{0.56}O_2$ cubic phase *Fm3m*, while Fe_2O_3 or other iron oxide phases are not visible (Figure 2(a)). XRD peaks are broad and the values of crystallite size obtained according to Scherrer equation are about 7.5 nm for sample CF and 3.5 nm for sample CZF. In order to understand better the structural properties of

Fe-doped system, CF and CZF catalysts were calcined at higher temperatures (1023 K).

After calcination, in the XRD profile of CF, peaks assigned to rhombohedral Fe_2O_3 (hematite) with R-$3c$ symmetry are visible (Figure 2(b)).

The lack of peak due to iron oxide in fresh CF samples could indicate the formation of solid solution between Ce and Fe. However, a comparison between lattice parameters retrieved from Rietveld refinement and from Vegard law (values of cell parameter expected if all the iron contained were dissolved in the lattice) indicates that only a small percentage of iron is dissolved in ceria (Table 2).

After aging, the increase of cell parameter indicates a segregation of the iron eventually dissolved in the lattice with formation of weak signal due to crystalline Fe_2O_3. It is known that lower valence ions such as Fe^{3+} are extremely difficult to dissolve into the ceria lattice, especially when treating at high temperature [61]. Mutual dissolution of Ce and Fe into Fe_2O_3 and CeO_2 has been reported to exist in Fe-rich Ce/Fe mixed oxides prepared by coprecipitation [62].

For CZF, the value of cell parameter retrieved by Rietveld refinement is not in agreement with that computed from Vegard's law: the adding of a cation (Fe^{3+}) with ionic radius smaller than Ce^{4+} and Zr^{4+} should produce a decrease in cell volume in the case of a solid solution. Conversely, we observe a value higher than expected indicating that Fe_2O_3 is probably deposited on the surface. Moreover, iron could be present as interstitial and/or extralattice or amorphous interparticle iron. As in the case of pure ceria, we cannot exclude that a small fraction of Fe is dissolved within ceria-zirconia framework.

9.3.2 CATALYTIC ACTIVITY

We investigated the heterogeneous process on different ceria-based catalysts performing reactions at 343 K for 2 hours, without any pH correction of the leachate (pH 9). Preliminary tests were carried out in order to verify the activity of catalyst and/or H_2O_2. In absence of catalyst and H_2O_2 (Figure 3), the abatement of COD and TOC, due only to the thermal treatment at 343 K, is small, respectively 1% and 14%.

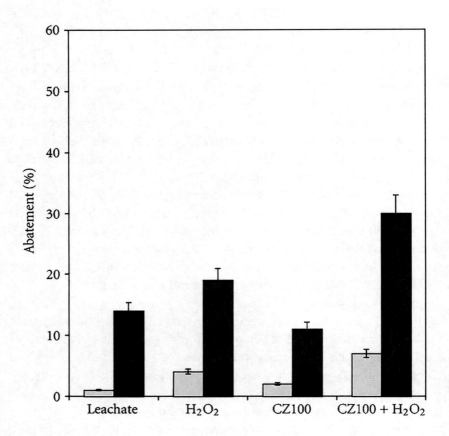

Figure 3. COD (light grey) and TOC (black) abatement for reaction with and without catalyst and H_2O_2 (reaction conditions: 10 mg of catalyst, 5 mL of H_2O_2, pH = 9, T = 343 K).

As shown in the plot, the advantage of the addition, for the abatement of COD and TOC, of bare ceria is negligible. In absence of the catalyst, but with 5 mL of H_2O_2 (3%), a small improvement in the activity was observed due to the oxidation capacity of the hydrogen peroxide alone. This activity could be explained considering that the small amount of iron presented in the leachate (Table 1) can interact with H_2O_2 (Fe/H_2O_2 ratio was 1 : 687)

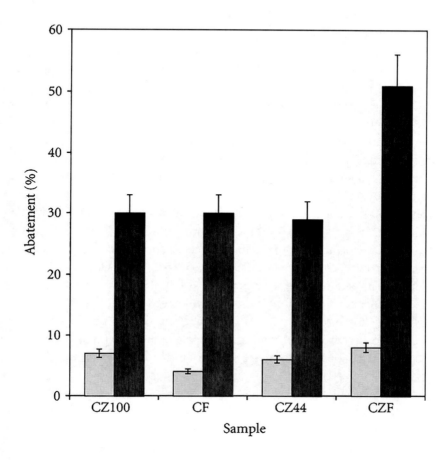

Figure 4. COD (light grey) and TOC (black) abatement for different catalysts (reaction conditions: 10 mg of catalyst, 5 mL of H_2O_2, pH = 9, T = 343 K).

catalyzing the formation of •OH radicals as in the homogeneous Fenton. When pure ceria and hydrogen peroxide were used in combination, the catalytic activity was further improved reaching an abatement of COD and TOC of 7% and 30%, respectively, confirming the positive synergic action of the two agents in the heterogeneous Fenton-like process. From these preliminary tests, we can conclude that ceria alone is not active and a synergic action between catalyst and hydrogen peroxide is necessary to obtain higher performance.

After blank tests, the activity of the four ceria-based catalysts (CZ100, CZ44, CF, and CZF) was investigated and the results are shown in Figure 4.

Ceria and ceria-zirconia solid solutions show very similar results. The catalytic activity of ceria-based systems could be attributed to the capacity of cerium oxide to decompose H_2O_2, as reported in a previous study in which the decomposition of hydrogen peroxide, with formation of radical species, in an aqueous suspension of CeO_2 was investigated [63]. The mechanism for H_2O_2 decomposition in the presence of water-oxide interfaces is still not completely elucidated, but it was suggested that it occurs on the surface with OH or HO_2 radicals production.

The catalytic activity of cerium oxide is correlated with its oxygen storage capacity. One of the most important roles of CeO_2 in catalytic redox reactions is to provide surface active sites [64] and to act as an oxygen storage/transport medium by its redox cycle between Ce^{4+} and Ce^{3+}. That is, the presence of surface active oxygens from one side and the oxygen storage capacity from the other are among the most important factors to be considered. These, in turn, are strongly influenced by surface area and surface/bulk composition.

As pointed out previously, doping ceria with Zr^{4+} increase, the oxygen mobility, but a higher amount of ZrO_2 decreases the number of redox sites and consequently the activity of the system. In order to explain the activity of the two systems, we need to take into account the right combination of surface area and composition.

For this reason, it is important to correlate overall activity with total available surface active oxygens, TSAO (which are linearly dependent on the amount of ceria) and total oxygen storage capacity, OSC (which generally shows a volcano-type relation with composition). The number of total surface oxygens (TSO) has been estimated according to Madier et al. [65] starting from the structure and the molar composition of the oxide considering the exposure of (100), (110), and (111) surfaces and assuming that Zr atoms do not participate in the redox process. The number of total surface available oxygens (TSAO) represents a fraction of total surface oxygens considering that only one atom out of four is involved in the Ce^{4+}-Ce^{3+} redox process [65–67]. OSC data collected according to the method described in the experimental. Results are reported in Table 3.

Even though CZ100 has a lower surface area, pure ceria and CZ44 show almost the same value of TSAO (225 µmol O/g and 182 µmol O/g, resp.). A more pronounced difference was found in OSC (1669 µg O_2/g and 3721 µg O_2/g for CZ100 and CZ44, resp.) that takes into account surface and bulk oxygens. In both catalysts, the surface area is quite high, consequently the abstraction of oxygen involves mainly surface sites, with little or no participation of the bulk in the reaction. Therefore, the more important factor is the availability of surface oxygen. In our materials (CZ100 and CZ44), the availability of surface oxygen is almost the same; therefore the two systems, CZ100 and CZ44, exhibit a very similar catalytic activity in the treatment of landfill leachate.

For CZF, the simultaneously presence of iron and zirconia significantly increased the abatement of TOC (51%) but has no significant effect on COD.

CF sample is characterized by the formation of cubic ceria-like solid solution where Fe cations are dissolved within ceria structure. In this case, the interaction takes place through the sharing of oxygen anion defined by the Fe–O–Ce bonds formed in the Fe-doped CeO_2 lattice [62].

In CZF sample, the lack of these interactions due to the lower amount of ceria and consequently to the lower amount of Ce-Fe-O entities formed in the system, can explain the different behavior of this catalyst. Indeed, in this case, a higher amount of Fe (due to weaker interaction with Ce and to the amorphous Fe_2O_3 phase on the surface) is available for the reaction with the leachate.

Our research pointed out the good activity of ceria-based heterogeneous treatment and we can conclude that ceria based catalyst is a very promising class of materials for this kind of application.

Further studies will be dedicated to a better understanding of the mechanism of reaction of ceria-based catalyst and to the optimization of the reaction conditions and catalytic stability.

Table 3. TSAO and OSC for CZ100 and CZ44.

Sample	TSAO (μmol O/g)	OSC (μg O$_2$/g)
CZ100	225	1669
CZ44	182	3721

9.4 CONCLUSIONS

Our study shows that the heterogeneous Fenton process could be successfully used in the treatment of landfill leachate substituting homogeneous treatment. Promising results were obtained in leachate oxidation by a heterogeneous Fenton-like process over ceria-based catalysts with an abatement of TOC higher than 50%.

This is just a first investigation into the potentiality of heterogeneous reaction, but the results appear encouraging. In heterogeneous reactions, several variables are involved and need to be completely understood for a good optimization of the catalyst. Further studies will be dedicated to a better understanding of the mechanism of reaction of ceria-based catalysts and the role of iron and zirconia in the reactions and leaching. Moreover, we need to optimize the reaction conditions, such as pH, temperature, and catalyst/peroxide ratio. Additional investigations should be performed in order to deeply explore a promising technique such as heterogeneous Fenton. At the moment, several aspects need to be investigated in more detail, but the results open a new field of research and point out a very interesting class of catalyst that could be used for landfill leachate treatment and worthy to be the subject of further investigations.

REFERENCES

1. L. H. Keith and W. A. Teliard, "Priority pollutants: a perspective view," Environmental Science & Technology, vol. 13, pp. 416–423, 1979.

2. K. Knox and P. H. Jones, "Complexation characteristics of sanitary landfill leachates," Water Research, vol. 13, no. 9, pp. 839–846, 1979.

3. M. Hagman, E. Heander, and J. L. C. Jansen, "Advanced oxidation of refractory organics in leachate—potential methods and evaluation of biodegradability of the remaining substrate," Environmental Technology, vol. 29, no. 9, pp. 941–946, 2008.

4. C. Papastavrou, D. Mantzavinos, and E. Diamadopoulos, "A comparative treatment of stabilized landfill leachate: coagulation and activated carbon adsorption vs. electrochemical oxidation," Environmental Technology, vol. 30, no. 14, pp. 1547–1553, 2009.

5. N. Meunier, P. Drogui, C. Montané, R. Hausler, G. Mercier, and J. F. Blais, "Comparison between electrocoagulation and chemical precipitation for metals removal from acidic soil leachate," Journal of Hazardous Materials, vol. 137, no. 1, pp. 581–590, 2006.

6. F. J. Rivas, F. Beltrán, F. Carvalho, B. Acedo, and O. Gimeno, "Stabilized leachates: sequential coagulation-flocculation plus chemical oxidation process," Journal of Hazardous Materials, vol. 116, no. 1-2, pp. 95–102, 2004.

7. M. J. K. Bashir, M. H. Isa, S. R. M. Kutty et al., "Landfill leachate treatment by electrochemical oxidation," Waste Management, vol. 29, no. 9, pp. 2534–2541, 2009.

8. E. Khan, R. W. Babcock, T. M. Hsu, and H. Lin, "Mineralization and biodegradability enhancement of low level p-nitrophenol in water using Fenton's reagent," Journal of Environmental Engineering, vol. 131, no. 2, pp. 327–331, 2005.

9. E. C. Catalkaya and F. Kargi, "Advanced oxidation of diuron by photo-fenton treatment as a function of operating parameters," Journal of Environmental Engineering, vol. 134, no. 12, pp. 1006–1013, 2008.

10. E. Neyens and J. Baeyens, "A review of classic Fenton's peroxidation as an advanced oxidation technique," Journal of Hazardous Materials, vol. 98, no. 1–3, pp. 33–50, 2003.

11. P. Westerhoff, G. Aiken, G. Amy, and J. Debroux, "Relationships between the structure of natural organic matter and its reactivity towards molecular ozone and hydroxyl radicals," Water Research, vol. 33, no. 10, pp. 2265–2276, 1999.

12. S. Zhao, H. Ma, M. Wang et al., "Role of primary reaction initiated by 254 nm UV light in the degradation of p-nitrophenol attacked by hydroxyl radicals," Photochemical and Photobiological Sciences, vol. 9, no. 5, pp. 710–715, 2010.

13. J. De Laat and H. Gallard, "Catalytic decomposition of hydrogen peroxide by Fe(III) in homogeneous aqueous solution: mechanism and kinetic modeling," Environmental Science and Technology, vol. 33, no. 16, pp. 2726–2732, 1999.

14. S. H. Lin and C. C. Lo, "Fenton process for treatment of desizing wastewater," Water Research, vol. 31, no. 8, pp. 2050–2056, 1997.

15. S. Parsons, Advanced Oxidation Processes for Water and Wastewater Treatment, IWA publishing, Alliance House, London, UK, 2004.

16. C. Jiang, S. Pang, F. Ouyang, J. Ma, and J. Jiang, "A new insight into Fenton and Fenton-like processes for water treatment," Journal of Hazardous Materials, vol. 174, no. 1–3, pp. 813–817, 2010.

17. M. D. Gurol and S. Lin, "Continuous catalytic oxidation processes," US PATENT 5755977, 1998.
18. S. Sabhi and J. Kiwi, "Degradation of 2,4-dichlorophenol by immobilized iron catalysts," Water Research, vol. 35, no. 8, pp. 1994–2002, 2001.
19. Y. F. Han, N. Phonthammachai, K. Ramesh, Z. Zhong, and T. I. M. White, "Removing organic compounds from aqueous medium via wet peroxidation by gold catalysts," Environmental Science and Technology, vol. 42, no. 3, pp. 908–912, 2008.
20. P. L. Huston and J. J. Pignatello, "Degradation of selected pesticide active ingredients and commercial formulations in water by the photo-assisted Fenton reaction," Water Research, vol. 33, no. 5, pp. 1238–1246, 1999.
21. Y. Deng and J. D. Englehardt, "Treatment of landfill leachate by the Fenton process," Water Research, vol. 40, no. 20, pp. 3683–3694, 2006.
22. A. Goi, Y. Veressinina, and M. Trapido, "Fenton process for landfill leachate treatment: evaluation of biodegradability and toxicity," Journal of Environmental Engineering, vol. 136, no. 1, pp. 46–53, 2010.
23. T. Yilmaz, A. Aygün, A. Berktay, and B. Nas, "Removal of COD and colour from young municipal landfill leachate by Fenton process," Environmental Technology, vol. 31, no. 14, pp. 1635–1640, 2010.
24. E. M. R. Rocha, V. J. P. Vilar, A. Fonseca, I. Saraiva, and R. A. R. Boaventura, "Landfill leachate treatment by solar-driven AOPs," Solar Energy, vol. 85, no. 1, pp. 46–56, 2011.
25. S. Mohajeri, H. A. Aziz, M. H. Isa, M. A. Zahed, and M. N. Adlan, "Statistical optimization of process parameters for landfill leachate treatment using electro-Fenton technique," Journal of Hazardous Materials, vol. 176, no. 1–3, pp. 749–758, 2010.
26. S. Navalon, R. Martin, M. Alvaro, and H. Garcia, "Gold on diamond nanoparticles as a highly efficient fenton catalyst," Angewandte Chemie—International Edition, vol. 49, no. 45, pp. 8403–8407, 2010.
27. R. Martín, S. Navalon, M. Alvaro, and H. Garcia, "Optimized water treatment by combining catalytic Fenton reaction using diamond supported gold and biological degradation," Applied Catalysis B, vol. 103, no. 1-2, pp. 246–252, 2011.
28. A. G. Chakinala, P. R. Gogate, A. E. Burgess, and D. H. Bremner, "Industrial wastewater treatment using hydrodynamic cavitation and heterogeneous advanced Fenton processing," Chemical Engineering Journal, vol. 152, no. 2-3, pp. 498–502, 2009.
29. I. Oller, S. Malato, and J. A. Sánchez-Pérez, "Combination of Advanced Oxidation Processes and biological treatments for wastewater decontamination-A review," Science of the Total Environment, vol. 409, pp. 4141–4166, 2010.
30. T. D. Nguyen, N. H. Phan, M. H. Do, and K. T. Ngo, "Magnetic Fe2MO4 (M:Fe, Mn) activated carbons: fabrication, characterization and heterogeneous Fenton oxidation of methyl orange," Journal of Hazardous Materials, vol. 185, no. 2-3, pp. 653–661, 2011.
31. N. Panda, H. Sahoo, and S. Mohapatra, "Decolourization of Methyl Orange using Fenton-like mesoporous Fe2O3-SiO2 composite," Journal of Hazardous Materials, vol. 185, no. 1, pp. 359–365, 2011.

Potential of Ceria-Based Catalysts for the Oxidation of Landfill Leachate 191

32. L. A. Galeano, M. Á. Vicente, and A. Gil, "Treatment of municipal leachate of landfill by fenton-like heterogeneous catalytic wet peroxide oxidation using an Al/Fe-pillared montmorillonite as active catalyst," Chemical Engineering Journal, vol. 178, pp. 146–153, 2011.
33. S. Bernal, J. Kaspar, and A. Trovarelli, "Recent progress in catalysis by ceria and related compounds—preface," Catal Today, vol. 50, pp. 173–173, 1999.
34. A. Trovarelli, Catalysis by Ceria and Related Materials, Imperial College Press, London, UK, 2002.
35. A. Trovarelli, C. De Leitenburg, M. Boaro, and G. Dolcetti, "The utilization of ceria in industrial catalysis," Catalysis Today, vol. 50, no. 2, pp. 353–367, 1999.
36. A. Trovarelli, C. De Leitenburg, and G. Dolcetti, "Design better cerium-based oxidation catalysts," Chemtech, vol. 27, no. 6, pp. 32–37, 1997.
37. L. Vivier and D. Duprez, "Ceria-based solid catalysts for organic chemistry," ChemSusChem, vol. 3, no. 6, pp. 654–678, 2010.
38. N. D. Tran, M. Besson, C. Descorme, K. Fajerwerg, and C. Louis, "Influence of the pretreatment conditions on the performances of CeO2-supported gold catalysts in the catalytic wet air oxidation of carboxylic acids," Catalysis Communications, vol. 16, no. 1, pp. 98–102, 2011.
39. S. Yang, W. Zhu, Z. Jiang, Z. Chen, and J. Wang, "nfluence of the structure of TiO2, CeO2, and CeO2-TiO2 supports on the activity of Ru catalysts in the catalytic wet air oxidation of acetic acid," Rare Metals, vol. 30, pp. 488–495, 2011.
40. J. J. Delgado, X. Chen, J. A. Pérez-Omil, J. M. Rodríguez-Izquierdo, and M. A. Cauqui, "The effect of reaction conditions on the apparent deactivation of Ce-Zr mixed oxides for the catalytic wet oxidation of phenol," Catalysis Today, vol. 180, pp. 25–33, 2011.
41. Y. Liu and D. Sun, "Effect of CeO2 doping on catalytic activity of Fe2O3/γ-Al2O3 catalyst for catalytic wet peroxide oxidation of azo dyes," Journal of Hazardous Materials, vol. 143, no. 1-2, pp. 448–454, 2007.
42. R. C. Martins, N. Amaral-Silva, and R. M. Quinta-Ferreira, "Ceria based solid catalysts for Fenton's depuration of phenolic wastewaters, biodegradability enhancement and toxicity removal," Applied Catalysis B, vol. 99, no. 1-2, pp. 135–144, 2010.
43. S. Silva Martínez, J. Vergara Sánchez, J. R. Moreno Estrada, and R. Flores Velásquez, "FeIII supported on ceria as effective catalyst for the heterogeneous photo-oxidation of basic orange 2 in aqueous solution with sunlight," Solar Energy Materials and Solar Cells, vol. 95, no. 8, pp. 2010–2017, 2011.
44. P. A. Deshpande, D. Jain, and G. Madras, "Kinetics and mechanism for dye degradation with ionic Pd-substituted ceria," Applied Catalysis A, vol. 395, no. 1-2, pp. 39–48, 2011.
45. APHA, AWWA, and WEF, Standard Methods for the Examination of Water and Wastewater, American Public Health Association, American Water Works Association, Water Environment Federation, Washington, DC, USA, 20th edition, 1999.
46. R. Jenkins and R. Snyder, To X-Ray Powder Diffractometry, Wiley, New York, NY, USA, 1996.
47. R. A. Young, The Rietveld Method IUCr, Oxford University Press, New York, NY, USA, 1993.

48. A. C. Larson and R. B. V. Dreele, General Structure Analysis System 'GSAS', Los Alamos National Laboratory, 2000.

49. B. H. Toby, "EXPGUI, a graphical user interface for GSAS," Journal of Applied Crystallography, vol. 34, no. 2, pp. 210–213, 2001.

50. D. J. Kim, "Lattice-parameters, ionic conductivities, and solubility limits in fluorite-structure MO_2 Oxide [M = Hf4+, Zr4+, Ce4+, Th4+, U4+] Solid Solutions," Journal of the American Ceramic Society, vol. 72, no. 8, pp. 1415–1421, 1989.

51. E. Aneggi, M. Boaro, C. De Leitenburg, G. Dolcetti, and A. Trovarelli, "Insights into the redox properties of ceria-based oxides and their implications in catalysis," Journal of Alloys and Compounds, vol. 408-412, pp. 1096–1102, 2006.

52. E. Mamontov, R. Brezny, M. Koranne, and T. Egami, "Nanoscale heterogeneities and oxygen storage capacity of Ce 0.5Zr0.5O2," Journal of Physical Chemistry B, vol. 107, no. 47, pp. 13007–13014, 2003. View at Google Scholar · View at Scopus

53. R. Q. Syed and W. Chiang, Sanitary Landfill Leachate, Generation, Control and Treatment, Technomic, Basel, Switzerland, 1994.

54. J. Kaspar and P. Fornasiero, "Structural properties and thermal stability of ceria-zirconia and related materials," in Catalysis by Ceria and Related Materials, A. Trovarelli, Ed., Imperial College Press, London, UK, 2002.

55. A. Trovarelli, F. Zamar, J. Llorca, C. De Leitenburg, G. Dolcetti, and J. T. Kiss, "Nanophase fluorite-structured CeO_2-ZrO_2 catalysts prepared by high-energy mechanical milling: analysis of low-temperature redox activity and oxygen storage capacity," Journal of Catalysis, vol. 169, no. 2, pp. 490–502, 1997.

56. S. Rossignol, F. Gérard, and D. Duprez, "Effect of the preparation method on the properties of zirconia-ceria materials," Journal of Materials Chemistry, vol. 9, no. 7, pp. 1615–1620, 1999.

57. H. Vidal, J. Kašpar, M. Pijolat et al., "Redox behavior of CeO_2-ZrO_2 mixed oxides. I. Influence of redox treatments on high surface area catalysts," Applied Catalysis B, vol. 27, no. 1, pp. 49–63, 2000.

58. H. Vidal, J. Kašpar, M. Pijolat et al., "Redox ehaviour of CeO_2-ZrO_2 mixed oxides II. Influence of redox treatments on low surface area catalysts," Applied Catalysis B, vol. 30, no. 1-2, pp. 75–85, 2001.

59. J. Kaspar, P. Fornasiero, G. Balducci, R. Di Monte, N. Hickey, and V. Sergo, "Effect of ZrO_2 content on textural and structural properties of CeO_2-ZrO_2 solid solutions made by citrate complexation route," Inorganica Chimica Acta, vol. 349, pp. 217–226, 2003.

60. P. Fornasiero, G. Balducci, R. Di Monte et al., "Modification of the redox behaviour of CeO_2 induced by structural doping with ZrO_2," Journal of Catalysis, vol. 164, no. 1, pp. 173–183, 1996.

61. Z. Tianshu, P. Hing, H. Huang, and J. Kilner, "Sintering and densification behavior of Mn-doped CeO_2," Materials Science and Engineering B, vol. 83, no. 1–3, pp. 235–241, 2001.

62. F. J. Pérez-Alonso, M. L. Granados, M. Ojeda et al., "Chemical structures of coprecipitated Fe-Ce mixed oxides," Chemistry of Materials, vol. 17, no. 9, pp. 2329–2339, 2005.

63. A. Hiroki and J. A. LaVerne, "Decomposition of hydrogen peroxide at water-ceramic oxide interfaces," Journal of Physical Chemistry B, vol. 109, no. 8, pp. 3364–3370, 2005.

64. A. Trovarelli, "Catalytic properties of ceria and CeO2-Containing materials," Catalysis Reviews, vol. 38, no. 4, pp. 439–520, 1996.

65. Y. Madier, C. Descorme, A. M. Le Govic, and D. Duprez, "Oxygen mobility in CeO2 and CexZr(1-x)O2 compounds: study by CO transient oxidation and 18O/16O isotopic exchange," Journal of Physical Chemistry B, vol. 103, no. 50, pp. 10999–11006, 1999.

66. M. Boaro, C. De Leitenburg, G. Dolcetti, and A. Trovarelli, "The dynamics of oxygen storage in ceria-zirconia model catalysts measured by CO oxidation under stationary and cycling feedstream compositions," Journal of Catalysis, vol. 193, no. 2, pp. 338–347, 2000.

67. C. E. Hori, H. Permana, K. Y. S. Ng et al., "Thermal stability of oxygen storage properties in a mixed CeO2-ZrO2 system," Applied Catalysis B, vol. 16, no. 2, pp. 105–117, 1998.

PART V

ASSESSMENT OF LEACHATE HAZARDS, PRE- AND POST-TREATMENT

CHAPTER 10

Investigation of Physicochemical Characteristics and Heavy Metal Distribution Profile in Groundwater System Around the Open Dump Site

S. KANMANI AND R. GANDHIMATHI

10.1 INTRODUCTION

The generation of solid waste has become an increasing environmental and public health problem everywhere in the world, particularly in developing countries. The fast expansion of urban, agricultural and industrial activities spurred by rapid population growth and the change in consumer habits has produced vast amounts of solid wastes (Akoteyon et al. 2011). Open dumps are the oldest and the most common way of disposing of solid waste. In recent years thousands have been closed, while many still are being used. In many cases, they are located wherever land is available, without regard to safety, health hazard and esthetic degradation. The waste is often piled as high as equipment allows. In some instances, the refuse is ignited and allowed to burn. In others, the refuse is periodically leveled and compacted (Sabahi et al. 2009). The dumping of solid waste

---

© *2013 by the authors.* Applied Water Science, *2013, 3:89, DOI: 10.1007/s13201-013-0089-y. Creative Commons Attribution license (http://creativecommons.org/licenses/by/3.0/).*

in uncontrolled landfills can cause significant impacts on the environment and human health (Dong et al. 2008). The most commonly reported danger to the human health from these landfills is from the use of groundwater that has been contaminated by leachate (Jhamnani and Singh 2009).

Leachate is produced when moisture enters the refuse in a landfill, extracts contaminants into the liquid phase, and produces moisture content sufficiently high to initiate liquid flow. Leachate is generated in a landfill as a consequence of the contact of water with solid waste (Lo 1996). Leachate from a solid waste disposal site is generally found to contain major elements like calcium, magnesium, potassium, nitrogen and ammonia, trace metals like iron, copper, manganese, chromium, nickel, lead and organic compounds like phenols, polyaromatic hydrocarbons, acetone, benzene, toluene, chloroform, etc. (Freeze and Cherry 1979). The concentration of these in the leachate and water depends on the composition of wastes (Alker et al. 1995). The rate and characteristics of leachate produced depends on many factors such as solid waste composition, particle size, degree of compaction, hydrology of site, age of landfill, moisture and temperature conditions, and available oxygen. Leachate tended to migrate in surrounding soil may result in contamination of underlying soil and groundwater (Jhamnani and Singh 2009). Leachate migrations from waste sites or landfills and the release of pollutants from sediment (under certain conditions) pose a high risk to groundwater resource if not adequately managed. Their impact on groundwater continues to raise concern and have become the subject of recent and past investigations (Ahmed and Sulaiman 2001; Fatta et al. 1999; Kjelsen et al. 1998; Bjerg et al. 1995; Robinson and Gronow 1992; Cariera and Masciopinto 1998; Loizidou and Kapetanios 1993; Gallorini et al. 1993; Khan et al. 1990; Kunkle and Shade 1976). Empirical investigations as well as modeling techniques (McCreanor and Reinhart 2000; Lee et al. 1997; Syriopoulou and Koussis 1987; Koussis et al. 1989; Ostendorf et al. 1984) have been used to assess the pollution of groundwater by leachate from a landfill.

Groundwater is an important source of drinking water for humankind. It contains over 90 % of the fresh water resources and is an important reserve of good quality water. Groundwater, like any other water resource, is not just of public health and economic value; it also has an important ecological function (Armon and Kitty 1994). The chemical composition of groundwater is a measure of its suitability as a source of water for human and animal consumption, irrigation, and for industrial and other purposes (Babiker et al. 2007). Therefore, monitoring the quality of water is important because clean water is necessary for human health and the reliability of aquatic ecosystems.

This study aims to develop an understanding of the natural groundwater quality in the Ariyamangalam open dumping site and the adjacent areas through the dug wells and bore wells that have been selected for this purpose. To estimate how far groundwater quality has been affected by the downward movement of leachate from the Ariyamangalam open dumping site, the surrounding groundwater samples were collected and analyzed for various physicochemical parameters. The heavy metals distribution profile in the groundwater system at the waste disposal site was also studied.

10.1.2 STUDY AREA

Tiruchirappalli, better called as Trichy, is the fourth largest municipal corporation in Tamil Nadu and also the fourth largest urban agglomeration in the state. Kaveri River flows through the length of the district and is the principal source of irrigation and water supply. The district has an area of 4,404 km^2. The district has an approximate population of 27 lakhs and the population density of 602 inhabitants per square kilometer. The maximum temperature experienced in this district is 37.7 °C and the minimum temperature is 18.9 °C. The normal annual rainfall is 842.60 mm. The Trichy district is served by an open dumping yard namely Ariyamangalam garbage ground located 12 km from the city on the Trichy–Thanjavur highway. The dumping site is located at 10°48′ N and 78°43′ E. The ground elevation of the dumping site is 75.875 m above mean sea level. The Ariyamangalam dumping site has been in operation since 1967, covering a total surface area of 47.7 acres where the geological formation consists of mainly alluvium (based on the data collected from Public Works Department, Tamilnadu). The Tiruchirappalli municipal corporation consists of four zones namely Srirangam, Ponmalai, Ariyamangalam, and Abishegapuram. The dump site receives approximately 400–470 tonnes of MSW per day, which are collected from four zones. The height of dump is around 3–4 m above ground level. The waste is disposed here without segregation and compaction. The deposited waste is not provided by any daily or intermediate covers. No lining is provided at the bottom of the dump yard to prevent leachate migration. Access to the site is freely available to all and the site is regularly utilized by rag pickers, scavenging for recyclable materials, and by a variety of animals including water buffaloes, cattle, pigs, and dogs scavenging food waste. Unhygienic drinking water and garbage-induced diseases, like dysentery, cholera and hepatitis, are frequently reported in the surrounding villages. The layout of the study area is shown in Fig. 1.

10.2 METHODOLOGY

10.2.1 SOLID WASTE COMPOSITION

To appraise the fresh solid waste composition, a total of 60 fresh solid waste samples (three samples from each zone per day) from four different zones of Trichy city were collected for 5 days from the truck discharge point to the dumping site. The collected fresh solid waste were segregated into different categories viz. paper, plastics, metals, glass, debris, vegetables, textiles, etc. The segregated samples were weighed individually (Mohan and Gandhimathi 2009). The mean of the solid waste composition was calculated using the results of the composition of each of the sorting samples (Gidarakos et al. 2005).

10.2.2 SAMPLING OF LEACHATE AND GROUNDWATER

Leachate samples from the actual leachate streams in the solid waste dumping site were collected in 5 L polypropylene carboys that were rinsed out thrice prior to sample collection, transported to the laboratory, stored at 4 °C and analyzed within 2 days. A total of 20 leachate samples were collected for monitoring purpose, out of which 11 samples were collected from the old dumping area (greater than 3 years old) referred to

Heavy Metal Distribution Profile in Groundwater System

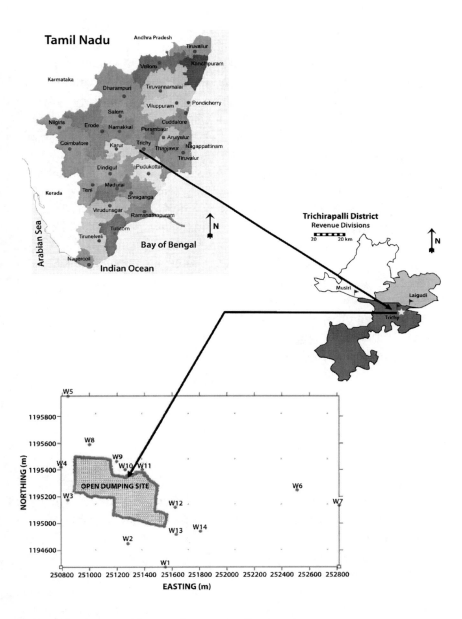

Figure 1. Study area with groundwater sampling locations.

as "stabilized leachate" whereas the remaining nine samples of leachate called as "fresh leachate" were collected near the new dumping area (less than or equal to 2 years old). The physicochemical analyses were carried out according to the standard methods (APHA 1998). A random sampling method was employed to collect the groundwater samples with due consideration to represent land-use patterns, topography, and areas close to dumping site (Kale et al. 2010). Groundwater samples of totally 14 numbers (4 hand pumps, 7 bore wells and 3 open wells) were collected within 1.7 km from the dumping site. The details of groundwater sampling locations are given in Table 1. The spatial data was obtained from the data recorded by a Garmin Global Positioning System. The groundwater sampling locations with the study area is shown in Fig. 1. Water was collected from open wells using drawing buckets tied with ropes, while bore wells and tube wells were pumped for 5–15 min before sampling (Andrew et al. 2011). These samples were collected in pre-cleaned polypropylene containers of 2 L capacity after rinsing with the sample and preserved airtight to avoid evaporation, stored at 4 °C and analyzed within 2 days.

The pH and electrical conductivity (EC) were recorded on site at the time of sampling with digital pH meter and digital EC meter, respectively. For heavy metal analyses, samples were separately collected in pre-washed polypropylene containers of 100 ml capacity and acidified onsite to avoid precipitation of metals.

10.2.3 ANALYTICAL METHODS

The parameters were selected based on their relative importance in municipal landfill leachates composition, and their pollution potential on groundwater resource in particular (Bagchi 2004). The physicochemical parameters such as total dissolved solids (TDS), total alkalinity (TA), total hardness (TH), major cations such as calcium (Ca^{2+}) and magnesium (Mg^{2+}), major anion such as chlorides (Cl^-) of the leachate and groundwater samples were analyzed by titrimetric methods. Chloride was included in the water quality assessment because of its measure of extent of dispersion of leachates in groundwater body (Chapman 1992). Sulfates (SO_4^{2-}) in the groundwater samples were analyzed by nephelometric turbidity

Table 1. Details of sampling wells.

Well No.	Type of well	Northing (m)	Easting (m)	Distance (km)
1	Hand pump	1,194,676	251,553	0.72
2	Open well	1,194,850	251,278	0.46
3	Bore well	1,195,175	250,840	0.38
4	Bore well	1,195,420	250,790	0.43
5	Hand pump	1,195,949	250,843	0.74
6	Bore well	1,195,248	252,510	1.31
7	Hand pump	1,195,133	252,816	1.62
8	Open well	1,195,590	251,000	0.35
9	Bore well	1,195,463	251,197	0.16
10	Open well	1,195,400	251,260	0.12
11	Bore well	1,195,401	251,384	0.21
12	Bore well	1,195,119	251,623	0.46
13	Hand pump	1,194,917	251,631	0.58
14	Bore well	1,194,940	251,806	0.70

method (APHA 1998). Nitrates ($NO_3{}^-$) and total organic carbon (TOC) determination in the groundwater samples were carried out by DR 2700 spectrometer. Estimation of chemical oxygen demand (COD) was done by closed reflux titrimetry method, while biochemical oxygen demand (BOD) was calculated by oxygen determination by Winkler titration for the preserved leachate sample. All the analyses in this study were repeated two or three times until concordant values were obtained, and all the tests were carried out according to the standard methods (APHA 1998). The data was statistically analyzed by setting up and calculating a correlation matrix for the various parameters using Statistical Package for Social Sciences (SPSS) software package (Norusis 1997).

The heavy metals such as Cd, Cu, Mn, Pb and Zn concentrations in the leachate and ground water samples were analyzed using atomic absorption spectrophotometer (AAS) supplied by Thermo Fisher Scientific, USA with D2 background correction lamp. Standard solutions of heavy metals viz. copper (Cu), cadmium (Cd), manganese (Mn), lead (Pb) and zinc (Zn) were prepared with distilled water using copper sulfate ($CuSO_4 \cdot 5H_2O$), cadmium sulfate ($CdSO_4 \cdot 8H_2O$), manganese sulfate ($MnSO_4 \cdot 7H_2O$), lead nitrate [$Pb(NO_3)_2$], and zinc nitrate [$Zn(NO_3)_2 \cdot 6H_2O$].

10.3 RESULTS AND DISCUSSION

10.3.1 COMPOSITION OF MUNICIPAL SOLID WASTE

The results from the composition studies (Fig. 2) show that samples from Ariyamangalam dumping site contained about 90–95 % combustible materials such as paper, textiles, plastic, debris, metals, glass, and vegetable waste. The non-combustible fraction such as metals and glass is about 1–5 %. The results were compared with solid waste composition reported by other researchers. Mohan and Gandhimathi (2009) reported that the MSW composition from Perungudi dumping site (Chennai City, Tamil Nadu) contained about 60–70 % combustible materials such as textile, leaves, plastics, food waste, etc., with an average of 65 %. The non-combustible fraction such as metals and glass was about 30–40 % with an average of 35 %. The solid waste composition in most Asian countries is highly biodegradable, mainly composed of an organic fraction with high moisture content (Visvanathan et al. 2004). Joseph et al. (2012) revealed that the biodegradable waste was high when compared to non-biodegradable waste in MSW composition at Chennai city. The findings from other studies indicated that the composition of MSW is site-specific and based on the nature of waste and source location.

10.3.2 LEACHATE

The physicochemical characteristics and the heavy metal contamination in the collected leachate sample depend primarily upon the waste composition and water content of the municipal solid waste (MSW) (Denutsui et al. 2012). The physicochemical characteristics and heavy metal

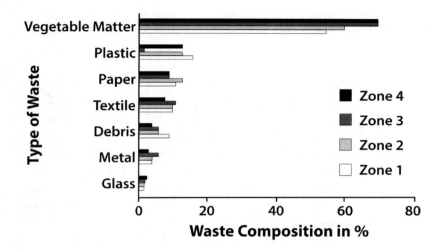

Figure 2. Fresh solid waste composition.

concentrations were analyzed for the collected leachate sample from Ariyamangalam open dumping site and are presented in Table 2.

10.3.3 BDL BELOW DETECTABLE LIMIT

From the Table 2, it can be observed that the fresh leachate sample possesses very high concentration of chemical parameters except pH, when compared to stabilized leachate samples. The pH value of the fresh samples was observed in the range of 6.96–7.77, but the older samples were amber colored and alkaline with pH range from 7.68 to 8.33. This may be attributed to the decrease in the concentration of free volatile acids due to anaerobic decomposition, as fatty acids can be partially ionized and contribute to higher pH values. Alkaline pH is normally encountered at landfills, 10 years after disposal (El-Fadel et al. 2002). The relatively

Table 2. Typical range of leachate characteristics.

S. No.	Parameter	Fresh leachate	Stabilized leachate
1	pH	6.05–7.77	7.68–8.33
2	Total dissolved solids	10,000–60,970	6,000–29,000
3	Electrical conductivity	11,130–57,000	6,810–40,300
4	Chlorides	1,000–8,000	1,000–6,500
5	Total alkalinity	4,500–26,000	2,500–11,000
6	Total hardness	4,000–17,000	4,500–8,500
7	Calcium	1,200–3,000	400–1,800
8	Magnesium	250–1,500	250–1,100
9	BOD_s	3,300–43,328	60–3,000
10	COD	6,240–75,840	1,024–19,200
11	Iron	14–600	40–400
12	Cadmium	0.35–1.40	BDL–0.16
13	Copper	0.85–1.92	BDL–0.55
14	Manganese	1.79–8.19	BDL–1.06
15	Lead	1.80–5.15	BDL–0.77
16	Zinc	0.35–4.80	BDL–0.19

All parameters are expressed in mg/L except pH and EC (EC is expressed in μ mho/cm)
BDL below detectable limit

high value of electrical conductivity (57,000 μ mho/cm) indicates the presence of dissolved inorganic materials in the samples. The concentration of TDS (60,970 mg/L) also fluctuates widely. Other inorganic contaminants also follow the trend of decreasing concentrations with increasing leachate age and stability. In general, leachate generated from young acidogenic landfills are characterized by high concentrations of organic and inorganic pollutants (Calli et al. 2005). The presence of Magnesium in the leachate is due to the disposal of construction waste along with MSW (Al-Yaqout 2003). The calcium and magnesium concentrations exhibited typical trends of constituents affected by the biological activity in the dumping

Heavy Metal Distribution Profile in Groundwater System

site. A reduction with time is attributed to the depletion of these compounds and to the increase in pH, thus reducing their solubility in leachate, and enhancing precipitation. The initial concentrations of these parameters are on the higher side (1,200–3,000 mg/L, 250–1,500 mg/L). A decrease in concentration was observed in stabilized leachate samples. The chloride concentration in the leachate varied from 1,000 to 8,000 mg/L for young leachate samples and 1,000–6,500 mg/L for stabilized samples. The possible anthropogenic sources of chloride are kitchen wastes from households, restaurants, and hotels. TA varies between 4,500 and 26,000 mg/L. The presence of high BOD (43,328 mg/L) and COD (75,840 mg/L) indicates the high organic strength. This indicates that majority of the organic compounds is biodegradable (Fatta et al. 1999). The presence of Fe (400 mg/L) in the leachate sample indicates that steel scraps are also dumped in the landfill. The dark brown color of the leachate is mainly attributed to the oxidation of ferrous to ferric form and the formation of ferric hydroxide colloids and complexes with fulvic/humic substance (Chu et al. 1994).

The concentration range of trace elements such as Cd, Cu, Fe, Mn, Pb, and Zn were found in the collected leachate samples and are also reported in Table 2. Fresh leachate samples showed a higher degree of metal solubilization, due to lower pH values caused by the biological production of organic fatty acids. As the dumping site age increased, the consequent increase in pH values caused a certain decrease in metal solubility (Mohan and Gandhimathi 2009). The stabilized leachate samples have less concentration when compared to fresh leachate samples. The concentration of Zn (4.80 mg/L) in the leachate shows that the dumping site receives waste from batteries and fluorescent lamps. The presence of Pb in the leachate samples are in the range 1.85–5.15 mg/L. The possible source of lead may be batteries, chemicals for photograph processing, older lead-based paints and lead pipes disposed at the landfill, which indicates toxicity to all forms of life at this level. Acidity in the leachate causes lead to be released from refuse (Al-Yaqout 2003). Cd (1.4 mg/L) and Cu (1.92 mg/L) are also present in the leachate samples whereas high Mn (8.18 mg/L) concentrations suggest a strong reducing environment. A variety of waste is dumped at Ariyamangalam open dump site, which likely indicate the origin of Mn, Pb, Cu, Cd, and Fe in leachate.

10.3.4 GROUNDWATER ASSESSMENT

The collected groundwater samples in and around the dumping site were free from color and odor excluding the locations W3 and W12. The groundwater of the studied area is used for drinking and domestic purposes. Table 3 shows the desirable and maximum permissible limit of individual species for drinking water recommended by Bureau of Indian Standards (BIS 1991) and World Health Organization (2002). The physicochemical concentrations of collected groundwater samples for the various parameters are shown in Figs. 3, 4, 5. The pH values of all groundwater samples are within the range of BIS and WHO standards. EC is the indicator of dissolved inorganic ions in groundwater. The presence of EC in the studied area ranges between 802 (W8) and 12,680 (W2) μ mho/cm and was found to be high, especially at locations W1, W2, W3, and W9. These high conductivity values measured in the underground water near the dumping site are an indication of its effect on groundwater. The TDS in groundwater reveal the saline behavior of water. According to classification by Rabinove et al. (1958), two samples were non saline (TDS value less than 1,000 mg/L); nine samples were in the slightly saline category (TDS value range between 1,000 and 3,000 mg/L) and two samples were in moderately saline category (TDS value range between 3,000 and 10,000 mg/L). Only one sample was very saline in nature (W2) with the value exceeding 10,000 mg/L. This high value of TDS may be due to the leaching of various pollutants into the groundwater.

Hardness is normally expressed as the total concentration of calcium and magnesium in mg/L, equivalent of $CaCO_3$. According to Durfor and Becker's (1964) classification of TH, only one groundwater sample (W7) was moderately hard among the 14 sampling locations. The remaining groundwater samples were very hard (greater than 180 mg/L as $CaCO_3$) in nature. Multivalent cations, particularly magnesium and calcium, are often present at a significant concentration in natural water. These ions are easily precipitated and in particular react with soap to make it difficult to remove scum. The calcium concentration in groundwater samples were beyond acceptable limit in two locations (W1 and W9). In most of the locations, magnesium content plays a major role and ranges between 7.2

Table 3. Drinking water quality standards as recommended by BIS and WHO.

S. No.	Parameters	BIS standards (IS 10500:1991)		World Health Organization (WHO) (2002)
		Desirable limit	Permissible limit	
1	pH	6.5–8.5	6.5–8.5	6.5–9.2
2	Total dissolved solids	300	1,500	250
3	Total alkalinity	200	600	500
4	Total hardness	300	600	300
5	Calcium	75	200	150
6	Magnesium	30	100	200
7	Chlorides	250	1,000	200
8	Sulfates	250	400	50
9	Nitrates	45	45	0.5
10	Cadmium	0.01	No relaxation	0.005
11	Copper	0.05	1.5	1
12	Iron	0.3	1.0	0.3
13	Manganese	0.1	0.3	0.1
14	Lead	0.05	No relaxation	0.05
15	Zinc	5	15	5

All values are in mg/L, except pH

and 81.6 mg/L. The TA was higher than the acceptable limit of 200 mg/L as $CaCO_3$ in all sampling locations. The high alkalinity imparts an unpleasant taste and may be deleterious to human health along with the high pH, TDS, and TH. Enhanced rock water interaction during post-monsoon could also contribute (to a limited extent) toward the increased values in bicarbonate (Pawar 1993).

The range of chlorides in all the locations under investigation is 215.15 (W8) to 4,098.73 (W2) mg/L. The concentration of chlorides in all locations except W8 exceeds the permissible level described by IS 10500-1991. Chloride in reasonable concentration is not harmful, but it causes corrosion in concentrations above 250 mg/L, while at about 400 mg/L, it causes a salty taste in water. An excess of chloride in water is usually

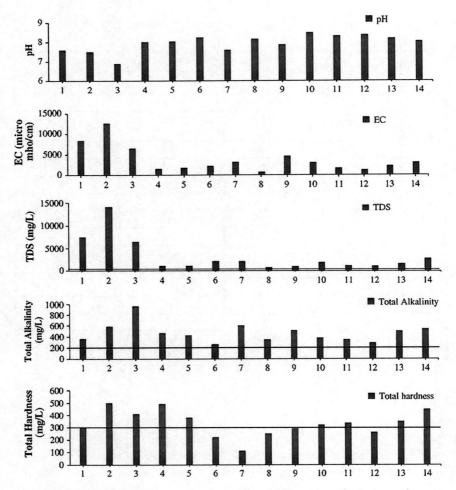

Figure 3. Concentrations of pH, EC, TDS, TA, and TH in groundwater samples.

taken as an index of pollution and considered as tracer for groundwater contamination (Loizidou and Kapetanios 1993).The chloride values in the water samples maybe due to the dissolution of rocks surrounding the aquifer and probably due to the leakage of sewage and anthropogenic pollution

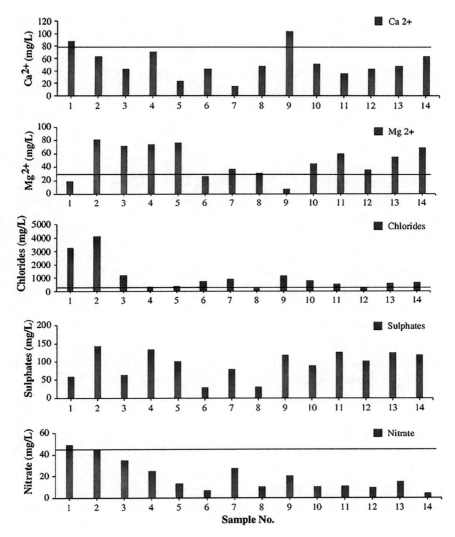

Figure 4. Concentrations of Ca^{2+}, Mg^{2+}, Cl^-, SO_4^{2-}, and NO_3^- in groundwater samples.

(agricultural activities). High concentration of chloride gives salty taste to water and may result in hypertension, osteoporosis, renal stones, and asthma (McCarthy 2004). The high chloride content in groundwater is from pollution sources such as domestic effluents, fertilizers, septic tanks, and leachates (Mor et al. 2006). Agricultural fertilizers and leachate are the main sources of sulfate in groundwater. The sulfate concentration in groundwater is within BIS and WHO standards for all the collected samples. Similarly, the nitrate concentration was also within the permissible limit (45 mg/L) in all the sampling locations except location 1 (W1). In general, the major sources for nitrate in groundwater include domestic sewage, runoff from agricultural fields, and leachate from landfill sites (Pawar and Shaikh 1995; Jawad et al. 1998; Lee et al. 2003; Jalali 2005). Higher concentration of NO_3^- in water causes a disease called "Methaemoglobinaemia" also known as "Blue-baby Syndrome". This disease particularly affects infants that are up to 6 months old (Kapil et al. 2009). The presence of TOC values ranges from 2.7 to 49 mg/L indicates that the groundwater contains organic impurities. The concentration of Fe in the groundwater samples varies from below detectable limit (BDL) to 5.102 mg/L (Fig. 5) and was found to be well above the WHO permissible limit (0.3 mg/L) in all the samples. Presence of Fe in water can lead to change of color of groundwater (Rowe et al. 1995).

The contour diagrams show (Figs. 6, 7, 8, 9, 10) the concentration profile of heavy metals at several water sample locations near the study area. The contour diagrams were drawn by surfer software. The collected groundwater samples were analyzed for heavy metals such as Cd, Cu, Mn, Pb, and Zn. Cadmium concentrations in the collected groundwater samples are lower than that of the leachate. The presence of cadmium concentration (Fig. 6) exceeds the permissible limit of 0.01 mg/L in all the collected samples. From the Fig. 6, it was observed that the contaminant species from the waste disposal site had migrated and accumulated at the location W6. Figure 7 shows the contour diagram of Cu distribution contaminant species from the location W3, it was located near an open dumping site toward southwest of the study area. From the Fig. 9, it was found that the presence of high concentration of Pb (0.59 mg/L) in groundwater samples nearby dumping site implies that groundwater samples were contaminated by leachate migration from an open dumping site. The concentrations of Zn (Fig. 10) in the collected samples were within the permissible limit.

Based on the contour diagram (from Figs. 6, 7, 8, 9, 10), the heavy metals such as lead (Pb), manganese (Mn), copper (Cu), Cadmium (Cd)

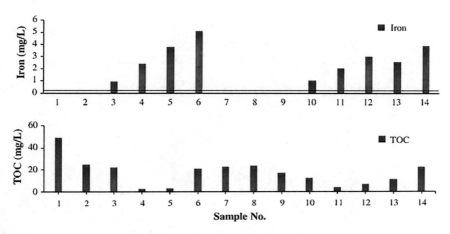

Figure 5. Concentrations of TOC and iron in groundwater samples.

showed significantly high values, which exceeded the maximum permissible concentration (MPC) as specified by WHO and BIS Standards for drinking water. Meanwhile, the zinc (Zn) was detected below the (MPC) value. The results indicated that all the collected sample locations were rigorously affected by the migration of leachate. Hence, the water is not potable for drinking purpose.

10.3.5 CORRELATION ANALYSIS FOR DIFFERENT WATER QUALITY PARAMETERS

Correlation is a method used to evaluate the degree of interrelation and association between two variables (Nair et al. 2005). A correlation of +1 indicates a perfect positive relationship between two variables. A correlation of −1 indicates that one variable changes inversely with relation to the other. A correlation of zero indicates that there is no relationship between the two variables (Kapil et al. 2009). Table 4 represents the correlation matrix among ten water quality parameters of groundwater of the study area.

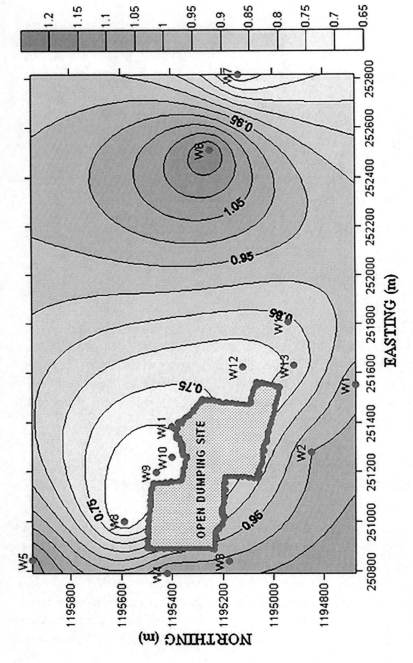

Figure 6. Contour diagram of cadmium distribution profile in groundwater samples.

Heavy Metal Distribution Profile in Groundwater System 215

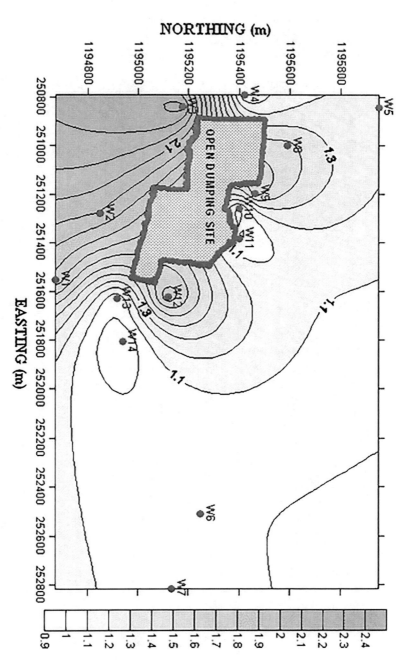

Figure 7. Contour diagram of copper distribution profile in groundwater samples.

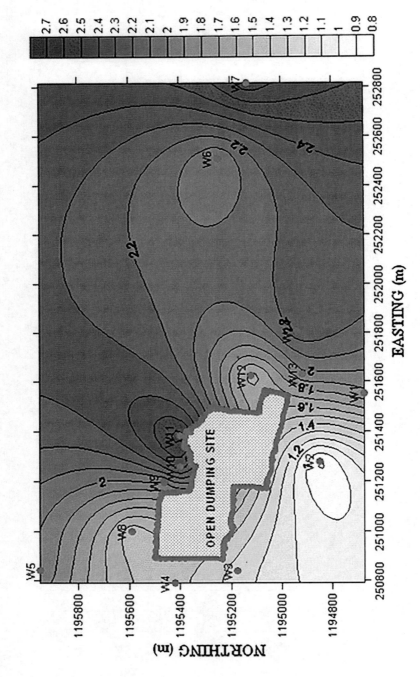

Figure 8. Contour diagram of manganese distribution profile in groundwater samples.

Heavy Metal Distribution Profile in Groundwater System 217

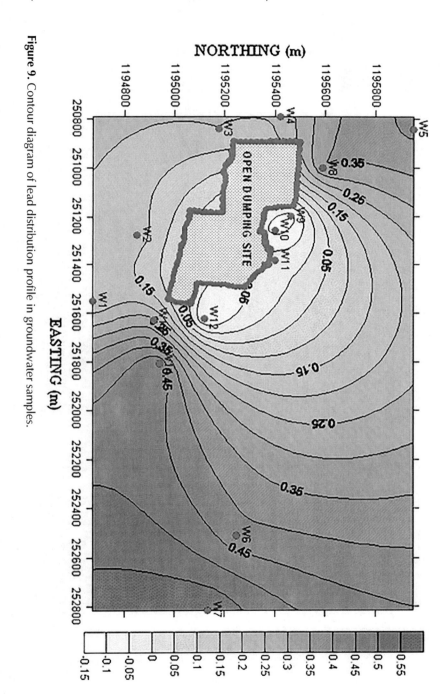

Figure 9. Contour diagram of lead distribution profile in groundwater samples.

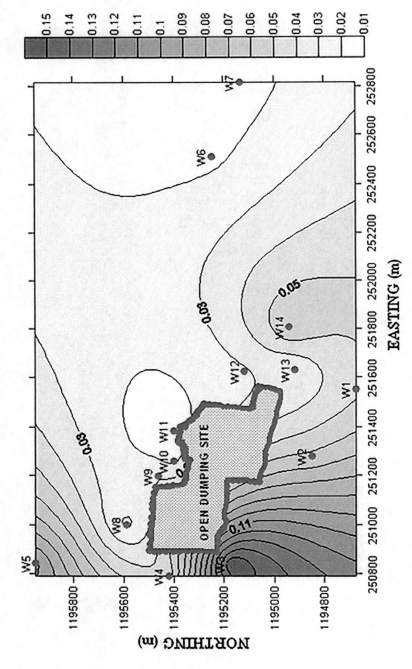

Figure 10. Contour diagram of zinc distribution profile in groundwater samples.

Table 4. Correlation matrix among ten water quality parameters of groundwater of study area.

	pH	EC	TDS	TH	TA	Ca^{2+}	Mg^{2+}	HCO$_3^-$	Cl$^-$	SO$_4^{2-}$	NO$_3^-$
pH	1										
EC	−0.679	1									
TDS	−0.750	0.748	1								
TH	−0.378	0.111	0.039	1							
TA	−0.838	0.428	0.476	0.425	1						
Ca^{2+}	−0.319	0.136	0.071	0.838	0.229	1					
Mg^{2+}	−0.332	0.061	0.001	0.882	0.485	0.482	1				
HCO$_3^-$	−0.864	0.431	0.477	0.432	0.989	0.242	0.486	1			
Cl$^-$	−0.535	0.964	0.704	0.077	0.201	0.157	−0.012	0.217	1		
SO$_4^{2-}$	0.085	0.165	−0.238	0.155	0.145	−0.032	0.276	0.139	0.123	1	
NO$_3^-$	−0.778	0.844	0.770	0.287	0.449	0.333	0.173	0.469	0.833	0.067	1

Electrical conductivity and Cl^- show good positive correlation with major water quality parameters. The correlation ($r = 0.964$, $p \leq 0.01$) between these two parameters for the analyzed samples in this study show a linear correlation. Equally, TA and HCO_3^- ($r = 0.989$, $p \leq 0.01$) indicate the good positive correlation. Some of the other highly significant and positive correlation were found between EC—NO_3^- ($r = 0.844$, $p \leq 0.01$), EC—TDS ($r = 0.748$, $p \leq 0.01$), TDS—HCO_3^- ($r = 0.477$, $p < 0.01$), TDS—Cl^- ($r = 0.704$, $p \leq 0.01$), TDS—NO_3^- ($r = 0.770$, $p \leq 0.01$), which also shows linear correlation. This suggested that presence of bicarbonate; chloride and nitrate in the study area greatly influence the TDS and EC. TH was also positively and significantly correlated with TA ($r = 0.425$, $p < 0.01$), HCO_3^- ($r = 0.432$, $p \leq 0.01$), SO_4^{2-} ($r = 0.155$, $p \leq 0.01$), Ca^{2+} ($r = 0.838$, $p \leq 0.01$) and Mg^{2+} ($r = 0.882$, $p \leq 0.01$). This result implies that there was great dependence of hardness on calcium, magnesium, chloride, sulfate, and bicarbonate. pH was negatively correlated with all the parameters except SO_4^{2-}. Similarly few other parameters were also found to have negative correlation, viz TDS—SO_4^{2-} ($r = -0.238$), Ca^{2+}—SO_4^{2-} ($r = -0.032$) and Mg^{2+}–Cl^- ($r = -0.012$). EC is solely a function of the major ion concentrations (Ca^{2+}, Mg^{2+}, Cl^-, SO_4^{2-}, and NO_3^-) of water quality parameters. The result was compared with the study reported by Kapil et al. (2009) for Indian conditions. They revealed that the major positive linear correlation was found in EC with all the water quality parameters in their study area.

10.4 CONCLUSION

The indiscriminate disposal and crude dumping of MSW are considered as a dangerous practice in integrated waste management at the global level. The fresh solid waste composition study shows that samples from the open dump site contained about 90–95 % combustible materials and non-combustible fraction is about 1–5 %. The fresh leachate sample possesses very high concentration of chemical parameters except pH, when compared to stabilized leachate samples. The heavy metal concentration range of collected fresh leachate sample showed a higher degree of metal solubilization when compared to stabilized leachate samples. The pH values of all groundwater samples are within the range of BIS and WHO standards. The physicochemical analysis indicated that chlorides (range between 215.15 and 4,098.73 mg/L) and TDS (ranges from 740 to 14,200 mg/L) of the groundwater samples are higher than the permissible limits in all the sampling locations when compared to other parameters. The presence of heavy metals (Pb, Mn, Cu, Cd and Zn) in groundwater samples indicates that there is appreciable contamination by leachate migration from an open dumping site. Based on the average concentration, the heavy metal components in the groundwater samples were found in the following order: Pb > Mn > Cu > Cd > Zn. From the present study, it was found that the groundwater is non-potable because most of the physicochemical parameters and heavy metals examined exceed the permissible limits. Ultimately, all results presented show that the Ariyamangalam open dump site constitutes a serious threat to local aquifers. The details such as solid waste composition, leachate characteristics and characteristics of groundwater samples will be used to develop contaminant transport model using Visual MODFLOW and MT3DMS software for leachate migration into subsurface system from open dumping site.

REFERENCES

1. Ahmed AM, Sulaiman WN (2001) Evaluation of groundwater and soil pollution in a landfill area using electrical resistivity imaging survey. Environ Manag 28:655–663
2. Akoteyon IS, Mbata UA, Olalude GA (2011) Investigation of heavy metal contamination in groundwater around landfill site in a typical sub-urban settlement in Alimosho Lagos-Nigeria. J Appl Sci Environ Sanit 6(2):155–163
3. Alker SC, Sarsby RW, Howell R (1995) Composition of leachate from waste disposal sites. In: Proceedings waste disposal by landfill—Green 1993, Balkemia, pp 215–221
4. Al-Yaqout AF (2003) Assessment and analysis of industrial liquid waste and sludge disposal at unlined landfill sites in arid climate. Waste Manag (Oxford) 23:817–824
5. Andrew AA, Jun S, Takahiro H, Kimpei I, George EN, Wilson YF, Gloria ETE, Ntankouo NR (2011) Evaluation of groundwater quality and its suitability for drinking domestic and agricultural uses in the Banana Plain (Mbanga Njombe Penja) of the Cameroon Volcanic Line. Environ Geochem Health 33:559–575e
6. APHA (1998) Standard methods for the examination of water and wastewater (17th ed). American Public Health Association, Washington

7. Armon R, Kitty (1994) The Health dimension of groundwater contamination. In: Holler (ed) Groundwater contamination and control. Marcel Dekker Inc, New York

8. Babiker SI, Mohamed AA, Mohamed TH (2007) Assessing groundwater quality using GIS. Water Resour Manag 21:699–715

9. Bagchi A (2004) Design of landfills and integrated solid waste management. Wiley, New Jersey

10. Bjerg PL, Rugge K, Pedersen JK, Christensen TH (1995) Distribution of redox-sensitive groundwater quality parameters downgradient of a landfill (Grindsted Denmark). Environ Sci Technol 29:1387–1394

11. Bureau of Indian Standards (BIS):1991 Indian standard specification for drinking water IS: 10500 2–4

12. Calli B, Mertoglu B, Inanc B (2005) Landfill leachate management in Istanbul: applications and alternatives. Chemosphere 59:819–829

13. Cariera C, Masciopinto C (1998) Assessment of groundwater after leachate release from landfills. Anal Chim 88:811–818

14. Chapman D (1992) A guide to the use of BIOTA sediments and water in environmental monitoring water quality assessments UNESCO/WHO/UNEP. Chapman and Hall, London, pp 371–460

15. Chu LM, Cheung KC, Wong MH (1994) Variations in the chemical properties of landfill leachate. Environ Manag 18:105–117

16. Denutsui D, Akiti TT, Osae S, Tutu AO, Blankson-Arthur S, Ayivor JE, Adu-Kwame FN, Egbi C (2012) Leachate characterization and assessment of unsaturated zone pollution near municipal solid waste landfill site at Oblogo Accra-Ghana. Res J Environ Earth Sci 4(1):134–141

17. Dong S, Liu B, Tang Z (2008) Investigation and modeling of the environment impact of landfill leachate on groundwater quality at Jiaxing Southern China. J Environ Technol Eng 1(1):23–30

18. Durfor CN, Becker E (1964) Public water supplies of the 100 largest cities in the United States. US Geographical Survey of Water Supply Paper, pp 1812

19. El-Fadel M, Bou-Zeid E, Chahine W, Alayli B (2002) Temporal variation of leachate quality from pre-sorted and baled municipal solid waste with high organic and moisture content. Waste Manag (Oxford) 22:269–282

20. Fatta D, Papadopoulos A, Loizidou M (1999) Evaluation of groundwater and soil pollution in a landfill area using electrical resistivity imaging survey. Environ Geochem Health 21:175–190

21. Freeze RA, Cherry JA (1979) Ground water. Prentice-Hall, Englewood Cliffs

22. Gallorini M, Pesavento M, Profumo A, Riolo C (1993) Analytical related problems in metal and trace elements determination in industrial waste landfill leachates. Sci Total Environ 133:285–298

23. Gidarakos E, Havas G, Ntzamilis P (2005) Municipal solid waste composition determination supporting the integrated solid waste management system in the island of Crete. Waste Manag 26(6):668–679

24. Jalali M (2005) Nitrate leaching from agricultural land in Hamadan western Iran. Agric Ecosyst Environ 110:210–218

25. Jawad A, Al-Shereideh SA, Abu-Rukah Y, Al Qadat K (1998) Aquifer ground water quality and flow in the Yarmouk River Basin of Northern Jordan. Environ Syst 26:265–287
26. Jhamnani B, Singh SK (2009) Groundwater contamination due to Bhalaswa landfill site in New Delhi. Int J Environ Sci Eng 1(3):121–125
27. Joseph Kurian, Rajendiran S, Senthilnathan R, Rakesh M (2012) Integrated approach to solid waste management in Chennai: an Indian metro city. J Mater Cycles Waste Manag 14:75–84
28. Kale SS, Ajay KK, Kumar Suyash, Pawar NJ (2010) Evaluating pollution potential of leachate from landfill site from the Pune metropolitan city and its impact on shallow basaltic aquifers. Environ Monit Assess 162:327–346
29. Kapil DM, Mamta K, Sharma DK (2009) Hydrochemical analysis of drinking water quality of Alwar District Rajasthan. Nature Sci 7(2):30–39
30. Khan R, Husain T, Khan HU, Khan SM, Hoda A (1990) Municipal solid waste management—a case study. Munic Eng 7:109–116
31. Kjelsen P, Bjerg PL, Rugge K, Christensen TH, Pedersen JK (1998) Characterization of an old municipal landfill (Grindsted Denmark) as a groundwater pollution source: landfill hydrology and leachate migration. Waste Manag Res 16:14–22
32. Koussis AD, Syriopoulou D, Ramanujam G (1989) Computation of three-dimensional advection-dominated transport in saturated aquifers. US Government Report
33. Kunkle GR, Shade JW (1976) Monitoring groundwater quality near a sanitary landfill. Groundwater 14:11–20
34. Lee KK, Kim YY, Chang HW, Chung SY (1997) Hydrogeological studies on the mechanical behaviour of landfill gases and leachate of the Nanjido Landfill in Seoul Korea. Environ Geol 31:185–198
35. Lee SM, Min KD, Woo NC, Kim YJ, Ahn C (2003) Statistical assessment of nitrate contamination in urban groundwater using GIS. Environ Geol 44:210–221
36. Lo IMC (1996) Characteristics and treatment of leachates from domestic landfills. Environ Int 22(4):433–442
37. Loizidou M, Kapetanios EG (1993) Effect of leachate from landfills on groundwater quality. Sci Total Environ 128:69–81
38. McCarthy MF (2004) Should we restrict chloride rather than sodium? Med Hypotheses 63:138–148
39. McCreanor PT, Reinhart DR (2000) Mathematical modeling of leachate routing in a leachate recirculating landfill. Water Res 34:1285–1295
40. Mohan S, Gandhimathi R (2009) Solid waste characterisation and the assessment of the effect of dumping site leachate on groundwater quality: a case study. Int J Environ Waste Manag 3(1/2):65–77
41. Mor S, Ravindra K, Dahiya RP, Chandra A (2006) Leachate characterization and assessment of groundwater pollution near municipal solid waste landfill site. Environ Monit Assess 118:435–456
42. Nair GA, Mohamed AI, Premkumar K (2005) Physico chemical parameters and correlation coefficents of ground waters of North–East Libya. Pollut Res 24(1):1–6

224 Sewage and Landfill Leachate

43. Norusis MJ (1997) SPSS Inc SPSS for Windows Professional Statistics 75. Prentice Hall, Englewood Cliffs
44. Ostendorf DW, Noss RR, Lederer DO (1984) Landfill leachate migration through shallow unconfined aquifers. Water Resour Res 20:291–296
45. Pawar NJ (1993) Geochemistry of carbonate precipitation from the ground waters in basaltic aquifers: an equilibrium thermodynamic approach. J Geol Soc India 41:119–131
46. Pawar NJ, Shaikh IJ (1995) Nitrate pollution of groundwater from shallow basaltic aquifers Deccan trap hydrologic province India. Environ Geol 25:197–204
47. Rabinove CJ, Long Ford RH, BrookHart JW (1958) Saline water resource of North Dakota. US geological survey of water supply paper 1428:72
48. Robinson H, Gronow J (1992) Groundwater protection in the UK: assessment of the landfill leachate source-term. Institute of Water Engineers and Managers 6:229-236
49. Rowe RK, Quigley RQ, Booker JR (1995) Clay barrier systems for waste disposal facilities. E and FN Spon, London
50. Sabahi AE, Abdul Rahim S, Wan Zuhairi WY, Fadhl AN, Fares A (2009) The characteristics of leachate and groundwater pollution at municipal solid waste landfill of Ibb City Yemen American. J Environ Sci 5(3):256–266
51. Syriopoulou D, Koussis AD (1987) Two-dimensional modeling of advection dominated solute transport in groundwater. Hydrosoft 1:63–70
52. Visvanathan C, Trankler J, Joseph K, Chiemchaisri C, Basnayake BFA, Gongming Z (2004) Municipal solid waste management in Asia. Asian Regional Research Program on Environmental Technology (ARRPET). Asian Institute of Technology publications. ISBN: 974:417-258-1
53. World Health Organization (WHO) (2002) Guideline for drinking water quality. Health criteria and other supporting information. World Health Organization, Geneva, pp 940–949

CHAPTER 11

Application of Response Surface Methodology (RSM) for Optimization of Semi-Aerobic Landfill Leachate Treatment Using Ozone

SALEM S. ABU AMR, HAMIDI ABDUL AZIZ, AND MOHAMMED J. K. BASHIR

11.1 INTRODUCTION

Sanitary landfill is recognized as the most common and desirable method for eliminating urban solid waste. It is also considered as the most economical and environmentally acceptable method for eliminating and disposing of municipal and industrial solid wastes (Tengrui et al. 2007). However, sanitary landfill generates a large amount of heavily polluted leachate (Zazouil and Yousefi 2008). The generation of leachate is mainly caused by a release from waste due to successive biological, chemical, and physical processes of waste deposited in a landfill. The quality and quantity of the water formed at landfills depend on several factors, including seasonal

---

© 2014 by the authors; licensee Springer. Applied Water Science, 2014, 4:156, DOI: 10.1007/s13201-014-0156-z. Creative Commons Attribution license (http://creativecommons.org/licenses/by/3.0/). Used with the authors' permission.

weather variations, land filling technique, phase sequencing, piling, and compaction method (Amonkrane et al. 1997; Trebouet et al. 2001).

Landfill leachate is a high-strength wastewater that is very difficult to deal with. Leachate generated from mature landfills (age >10 years) is typically characterized by large amounts of organic contaminants measured as chemical oxygen demand (COD), biochemical oxygen demand (BOD_5), ammonia, halogenated hydrocarbons suspended solid, significant concentration of heavy metals, and many other hazardous chemicals identified as potential sources of ground and surface water contamination (Schrab et al. 1993; Christensen et al. 2001; Renou et al. 2008; Aziz et al. 2009; Foul et al. 2009). Moreover, the sequent migration of leachate away from landfill and its release into the environment are serious environmental pollution concerns, threatening public health and safety (Read et al. 2001). Accordingly, many environmental specialists are determined to find efficient treatments for large quantities of polluted leachate.

A number of leachate treatment techniques have been applied, which include biological, physical, and chemical processes (Baig and Liechti 2001; Goi et al. 2009). Given the oxidation efficiency, ozone has been suggested as one of the chemical processes used for the treatment of stabilized landfill leachate to reduce the risk of strength and un-biodegradable organics (Beaman et al. 1998). Ozonation processes are effective means for the treatment of landfill leachates due to the high oxidative power of ozone (Huang et al. 1993; Rice 1997; Haapea et al. 2002; Wu et al. 2004). During ozonation, the biodegradability of leachate will be enhanced due to the fragmentation of organic compounds with long chains to lower chains degraded to carbon dioxide (Geenens et al. 2001). The performance of ozone for removing COD and color from mature landfill leachate has been demonstrated in the literature (Rivas et al. 2003; Chaturapruek et al. 2005; Hagman et al. 2008; Goi et al. 2009; Cortez et al. 2011). However, none of these reports have evaluated the effects of different O_3 dosages for different concentrations of leachate during different reaction times.

In the present study, the statistical relationships among three independent factors (ozone dosage, COD concentration, and reaction time) for the treatment of semi-aerobic stabilized leachate were assessed through RSM. The RSM is a mathematical and statistical technique that is useful for the optimization of chemical reactions and industrial processes and is

Optimization of Semi-Aerobic Landfill Leachate Treatment Using Ozone 227

commonly used for experimental designs. The main objectives of the present study include the following:

1. To investigate the efficiency of ozone for treating semi-aerobic stabilized leachate with different concentration levels.
2. To build up the equations of COD, ammoniacal nitrogen, and color removal efficiency from stabilized leachate and ozone consumption with respect to operational conditions [i.e., ozone dosage, reaction time, and COD concentration using RSM and central composite design (CCD)].
3. To determine the optimum operational condition of the studied application.

11. 2 MATERIALS AND METHODS

11.2.1 LEACHATE SAMPLING AND CHARACTERISTICS

The leachate samples used in the current study were collected from the aeration pond of a semi-aerobic stabilized leachate of the Pulau Burung landfill site (PBLS) in Nibong Tebal, Penang, Malaysia. The total landfill site area is 62.4 ha; however, only 33 ha are currently utilized to receive about 2,200 tons of solid waste daily (Bashir et al. 2011). This landfill produces a dark-colored liquid with pH level of more than 7.0, and is classified as stabilized leachate with high concentration of COD, NH_3–N, and low BOD/COD ratio (Aziz et al. 2007). All samples were collected manually in 20 l plastic containers, and then transferred, characterized, and refrigerated immediately in accordance with the Standard Methods for the Examination of Water and Wastewater (APHA 2005). Table 1 shows some characteristics of the leachate sample.

11.2.2 EXPERIMENTAL PROCEDURES

All experiments were carried out in a 2-L volume of sample using an ozone reactor with a height of 65 cm and an inner diameter of 16.5 cm

228   Sewage and Landfill Leachate

Table 1. Characteristics of semi aerobic landfill leachate from PBLS.

Parameters	Value
COD (mg/l)	2,000
NH3–N (mg/l)	960
Color (PT Co.)	3,670
pH	8.5
Suspended solids (mg/l)	197
Conductivity (µS/cm)	16,650

and supported by a cross-column ozone chamber for enhancing ozone gas diffusion (Fig. 1). Ozone was produced by a BMT 803 generator (BMT Messtechnik, Germany) fed with pure dry oxygen with recommended Gas flow rate of 100–1,000 ml/min under 1 bar pressure. Gas ozone concentration (in g/m^3 NTP) was measured by an ultraviolet gas ozone analyzer (BMT 964). The water bath and cooling system supported the ozone reactor to keep the internal reaction temperature at <15 °C. The process variables include ozone dosage, reaction time, and varied COD concentrations of leachate. Concentrations of COD, color, and ammonia were tested before and after each ozonation process, and the removal efficiency was then conducted. All tests were conducted according to the standard methods for the examination of water and wastewater (American Public Health Association (APHA) 2005). Ozone consumption (OC) in removing a certain amount of COD during ozonation under experimental conditions is given in the following Eq. (1):

$$OC = OC = \frac{Q_G}{V} \times \frac{\int_0^t \left(1 - \frac{C_{AG}}{C_{AG0}}\right) dt}{(COD_0 - COD)} \tag{1}$$

where Q_G is the gas flow rate (ml/min); V is the sample volume (ml); C_{AG} is the off-gas ozone concentration (g/m^3); C_{AG0} is the input ozone concentration (g/m^3); t is the time (min); and COD_0 and COD correspond to the initial and final COD (mg/l).

Optimization of Semi-Aerobic Landfill Leachate Treatment Using Ozone

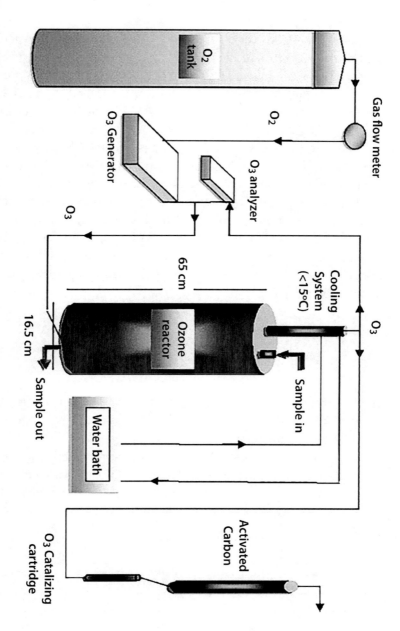

Figure 1. Schematic diagram of ozone equipment

11.2.3 EXPERIMENTAL DESIGN AND ANALYSIS

The Design Expert Software (version 6.0.7) was used for the statistical design of experiments and data analysis. In the present study, the CCD and response surface methodology (RSM) were applied to optimize and assess the relationship among three significant independent variables: (1) ozone dosage, (2) reaction time, and (3) COD concentrations in leachate as presented in Table 2. COD, color, and NH_3–N removal were considered as the dependent factors (response). Performance of the process was evaluated by analyzing the COD, color, and NH_3–N removal efficiencies. Each independent variable was varied over three levels between -1 and $+1$ at the determined ranges based on a set of preliminary experiments. The total number of experiments obtained for the three factors was 20 (=2 k $+ 2^k + 6$), where k is the number of factors (k = 3). Fourteen experiments were enhanced with 6 replications to assess the pure error. Considering that there are only three levels for each factor, the appropriate model is the quadratic model Eq. (2).

$$Y = \beta_0 + \sum_{j=1}^{k} \beta_j X_j + \sum_{j=1}^{k} \beta_{jj} X_j^2 + \sum_{i} \sum_{<j=2}^{k} \beta_{ij} X_i X_j + e_i \qquad (2)$$

where Y is the response; X_i and X_j are the variables; β is the regression coefficient; k is the number of factors studied and optimized in the experiment; and e is the random error.

Analysis of variance (ANOVA) was used for graphical analyses of the data to obtain the interaction between the process variables and the

Table 2. Independent variables of the CCD design.

Level of value	Ozone dosage (g/m³)	COD concentration (mg/l)	Reaction time (min)
-1	30	250	10
0	55	1,125	35
$+1$	80	2,000	60

responses. The quality of the fit polynomial model was expressed by the value of correlation coefficient (R^2), and its statistical significance was checked by the F test in the same program. Model terms were evaluated by the P value (probability) with 95 % confidence level.

11.3 RESULTS AND DISCUSSION

There were a total of 20 runs of the CCD experimental design, and the results are shown in Table 3. The observed percent removal efficiencies varied between 4 and 27.2 % for COD, 0–8.5 % for NH_3–N, and 11–90 % for color. Several researchers have conducted studies on the treatment of mature landfill leachate using ozone. Tizaoui et al. (2007) obtained 27 and 87 % removal for COD and color, respectively, after 60 min Ozonation of raw leachate. In the same way, Hagman et al. (2008) obtained 22 % COD reduction. Rivas et al. (2003) obtained a 30 % depletion of COD. Accordingly, the efficiency of ozone technique for solely removing organics and ammonia from leachate is relatively weak; the technique is more efficient for color removal, which may be attributed to the strength of organic components in leachate, improving the removal efficiency in lower initial COD concentration as shown in Table 3. Thus, many researchers have employed several advanced oxidation agents and techniques to improve the efficiency of ozone for leachate treatment, such as hydrogen peroxide (H_2O_2) and UV (Wu et al. 2004; Tizaoui et al. 2007). Other experiments have used lower pH and adsorbent materials, such as activated carbon, to enhance the removal of ammonia from leachate during ozonation (Park and Jin 2005).

Ozone consumption was also calculated under conditions of each run by following Eq. (1), and ranged from 1.6 to 19.40 (kgO_3/kg COD). OC is defined as the amount of ozone gas consumed for removing a certain amount of COD during ozonation under experimental conditions. OC value increased at minimum reaction time (10 min) and maximum initial COD concentration (2,000 mg/l). This result suggests that ozone running at maximum time (60 min) will reduce the amount of OC compared with an improved COD removal efficiency. Several experiments with ozone consumption values have been conducted from less than 1 kgO_3/kg COD

232 Sewage and Landfill Leachate

Table 3. Response values for different experimental conditions.

Run no.	Factor A	Factor B	Factor C	Response 1	Response 2	Response 3	Response 4
	Ozone dosage (g/m³)	COD concentration (mg/l)	Reaction time (min)	COD removal (%)	NH3–N removal (%)	Color removal (%)	OC (kgO₃/ kg COD)
1	80	250	60	27.2	8.5	90	19.40
2	55	1,125	35	18.8	1.1	31.8	3.62
3	30	250	10	16	0.0	25	3.44
4	55	2,000	35	21	0.0	24	1.80
5	80	2,000	10	10	0.0	18.5	2.04
6	55	250	35	24	6.5	72	7.72
7	55	1,125	35	17.5	1.2	32.5	3.41
8	80	2,000	60	15	0.0	27.3	6.96
9	55	1,125	35	18	1.1	33	3.33
10	55	1,125	35	18.5	1.2	32	3.41
11	30	2,000	10	4	0.0	11	1.80
12	55	1,125	35	17	0.9	31	3.70
13	30	2,000	60	11	0.0	23	2.09
14	55	1,125	10	15.5	0.0	16	1.60
15	30	1,125	35	12.5	1.0	38	5.15
16	80	250	10	15	1.0	38	9.47
17	55	1,125	35	17.5	1.2	33.6	4.50
18	30	250	60	20.8	2.0	88	4.72
19	55	1,125	60	22	1.4	58	3.18
20	80	1,125	35	19	2.2.	31	6.09

(Ho et al. 1974), 0.63 kgO₃/kg COD (Abu Amr and Aziz 2012) 3.5 kgO₃/ kg COD (Tizaoui et al. 2007), up to 16 kgO₃/kg COD (Wang et al. 2003), and between 2 and 3 for ozone alone systems (Geissen 2005).

11.3.1 ANALYSIS OF VARIANCE

Table 4 present the ANOVA of regression parameters of the predicted response surface quadratic models and other statistical parameters for COD,

Optimization of Semi-Aerobic Landfill Leachate Treatment Using Ozone 233

NH3–N, color removal, and OC. Data given in these tables demonstrate that all the models were significant at the 5 % confidence level, given that P values were less than 0.05. The values of correlation coefficient (R^2 = 0.8468, 0.9439, 0.9536, and 0.8949) obtained in the present study for COD, NH_3–N, color removal, and OC were higher than 0.80. For a good fit of model, the correlation coefficient should be at a minimum of 0.80. A high R^2 value close to 1 illustrates good agreement between the calculated and observed results within the range of experiment and shows that a desirable and reasonable agreement with adjusted R^2 is necessary (Joglekar and May 1987; Nordin et al. 2004). The "Adequate Precision" ratio of the models varies between 16.214 and 28.772, which is an adequate signal for the model. AP values higher than 4 are desirable and confirm that the predicted models can be used to navigate the space defined by the CCD.

COD: SD = 2.27, PRESS = 148.07, R^2 = 0.8468, Adj R^2 = 0.8060, Adeq precision = 17.508
NH3–N: SD = 0.69, PRESS = 24.12, R^2 = 0.9439, Adj R^2 = 0.9111, Adeq precision = 19.710
Color: SD = 5.38, PRESS = 1,053.30, R^2 = 0.9536, Adj R^2 = 0.9412, Adeq Precision = 28.772
OC: SD = 2.91, PRESS = 236.61, R^2 = 0.5289, Adj R^2 = 0.4735, Adeq Precision = 10.072

In the current study, four quadratic models are significant model terms (Table 4). Insignificant model terms, which have limited influence, were excluded from the study to improve the models. Based on the results, the response surface models constructed for predicting COD, NH_3–N, color removal efficiency, and OC were considered reasonable.

The final regression models, in terms of their coded and actual factors, are presented in Table 5. To confirm if the selected model provides an adequate approximation of the real system, the normal probability plots of the studentized residuals and diagnostics are provided by the Design Expert 6.0.7 software. The normal probability plots that helped us judge the models (Fig. 2a–c) demonstrate the normal probability plots of the standardized residuals for COD, NH_3–N, color removal, and OC. A normal probability plot indicates that if the residuals follow a normal distribution, as shown in Fig. 1, the points will follow a straight line for each case. However, some scattering is expected even with the normal data. Accordingly, the data can be possibly considered as normally distributed in the responses of certain models.

234   Sewage and Landfill Leachate

Table 4. ANOVA for analysis of variance and adequacy of the quadratic model for COD, NH_3–N, and Color removal and OC

	Source	Sum of squares	Degree of freedom	Mean square	F value	Prob > F
COD	Model	427.61	4	106.90	20.73	<0.0001
	A	47.96	1	47.96	9.30	0.0081
	B	176.40	1	176.40	34.21	<0.0001
	C	126.03	1	126.03	24.44	0.0002
	A^2	77.22	1	77.22	14.97	0.0015
	Residual	77.35	15	5.16		
	Lack of fit	75.05	10	7.50	16.26	0.0033
	Pure error	2.31	5	0.46		
NH_3–N	Model	95.79	7	13.68	28.82	<0.0001
	A	15.38	1	15.38	32.38	0.0001
	B	47.09	1	47.09	99.17	<0.0001
	C	5.18	1	5.18	10.92	0.0063
	B^2	8.26	1	8.26	17.39	0.0013
	C^2	2.85	1	2.85	6.00	0.0306
	AB	15.68	1	15.68	33.02	<0.0001
	BC	4.20	1	4.20	8.86	0.0116
	Residual	5.70	12	0.47		
	Lack of fit	5.63	7	0.80	58.85	0.0002
	Pure error	0.068	5	0.014		
	Model	95.79	7	13.68	28.82	<0.0001
Color	Model	8,919.57	4	2,229.89	77.04	<0.0001
	B	4,674.24	1	2,917.26	100.78	<0.0001
	C	2,917.26	1	377.58	13.04	<0.0001
	B^2	377.58	1	950.48	32.84	0.0026
	BC	950.48	1	950.48	32.84	< 0.0001
	Residual	434.20	15	28.95		
	Lack of fit	429.95	10	42.99	50.60	0.0002
	Pure error	4.25	5	0.85		
OC	Model	161.97	2	80.99	9.54	<0.0017
	A	71.61	1	71.61	8.44	0.0099
	B	90.36	1	90.36	10.65	0.0046
	Residual	144.27	17	8.49		
	Lack of fit	143.33	12	11.94	63.35	<0.0001
	Pure error	0.94	5	0.19		

Table 5. Final equations in terms of coded and actual factors for parameters.

	Final equation in terms of coded factors	Final equation in terms of actual factors
COD removal (%)	$+18.98 + 2.19\ A - 4.20\ B + 3.55\ C - 3.93\ A^2$	$-4.42920 + 0.77928\ A - 4.80000E{-}003\ B + 0.14200\ C + 6.28800E \\ - 003\ A^2$
NH_3–N removal (%)	$+1.32 + 1.24\ A - 2.17\ B + 0.72C + 1.61\ B^2 - \\ 0.94C^2 - 1.40\ A\ B - 0.72\ BC$	$-4.08677 + 0.12160\ A - 2.52041E{-}003\ B + 0.17179\ C - \\ 2.09796E{-}006\ B^2 - 1.51000E{-}003\ C^2 - 6.40000E{-}005\ A\ B - \\ 3.31429E{-}005\ BC$
Color removal (%)	$+33.69 - 21.62\ B + 17.08\ C + 8.69\ B^2 - 10.90 \\ BC$	$+32.32024 - 0.032807\ B + 1.24377\ C + 1.13502E{-}005\ B^2 - \\ 4.98286E{-}004\ BC$
OC (kgO_3/kg COD)	$+4.87 + 2.68\ A\ 3.01\ B$	$+2.84916 + 0.10704\ A\ 3.43543E{-}003\ C$

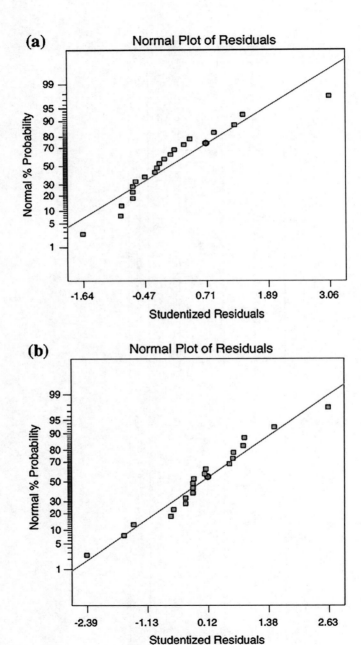

Figure 2. Design expert plot; normal probability plot of the standardized residual for **a** COD, **b** NH_3-N, **c** color removal and **d** OC

Optimization of Semi-Aerobic Landfill Leachate Treatment Using Ozone

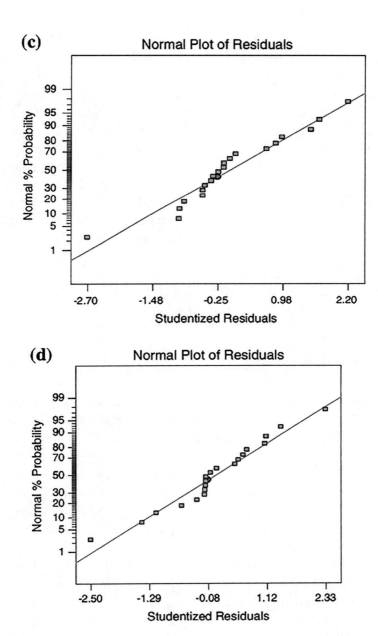

Figure 2. Design expert plot; normal probability plot of the standardized residual for **a** COD, **b** NH$_3$–N, **c** color removal and **d** OC

11.3.2 TREATMENT EFFICIENCY

To assess the interactive relationships between independent variables and the responses of certain models, the 3D surface response and contour plots utilized the Design Expert 6.0.7 software (Figs. 3, 4, 5, 6). As shown in Figs. 3a and 4a, the maximum observed removal of COD and NH_3–N were 27 and 8.2 %, respectively, at ozone dosage 80 g/m^3 and COD concentration 2,000 mg/l. The contour plots demonstrate that the improvement of removal efficiencies for COD and NH_3–N is attributed to the decrease in COD concentration and increase in ozone dosage (Figs. 3b, 4b). The maximum removal of color was 92 % at ozone dosage 80 g/m^3 and 60 min reaction time (Fig. 5a). The increase in reaction time and decrease in COD are the two main factors for improving color removal (Fig. 5b). Based on the target of OC as a minimum value, Fig. 6 shows the response and contour plot for the amount of ozone gas consumption for COD reduction-based cretin-independent variables; a minimum value of OC will follow the increase in reaction time and ozone dosage.

11.3.3 OPTIMIZATION PROCESS

The optimization process was carried out to determine the optimum value of COD, NH_3–N, and color removal efficiency, in addition to OC for COD removal using the Design Expert 6.0.7 software. According to the software optimization step, the desired goal for each operational condition (ozone dosage, COD concentration, reaction time) was chosen "within" the range. The responses (COD, NH_3–N, and color) were defined as maximum to achieve the highest performance, whereas the OC response was defined as the minimum to achieve the lowest value of ozone Gas consumed for removing the highest amount of COD. The program combines the individual desirabilities into a single number and then searches to optimize this function based on the response goal. Accordingly, the optimum working conditions and respective percent removal efficiencies were established, and the results are presented in Table 6. As shown in Table 6, 26.7, 7.1, and 92 % removal of COD, NH_3–N, and color are predicted, respectively, whereas

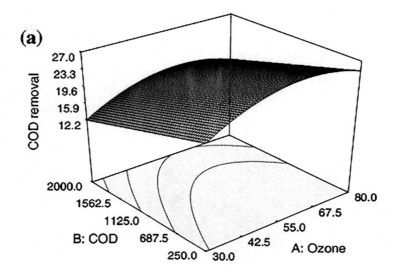
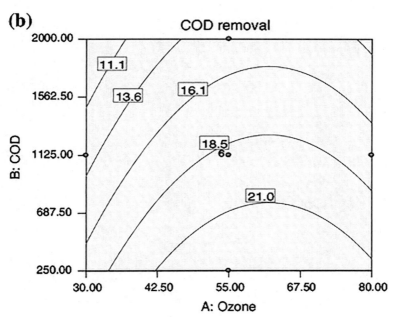

Figure 3. Response surface (a) and contour plots (b) for COD removal efficiency as a function of ozone dosage, (30 g/m3), COD concentration, (250 mg/l) and reaction time, (60) min

OC is presented as 9.40 (kgO$_3$/kg COD) based on the model under optimized operational conditions (ozone dosage 70 g/m^3; COD concentration 250 mg/l; and reaction time 60 min). The desirability function value was found to be 0.823 for these optimum conditions. An additional experiment was then performed to confirm the optimum results. The laboratory experiment agrees well with the predicted response value.

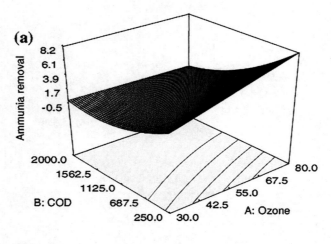

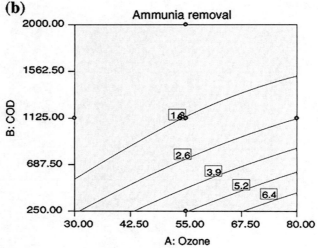

Figure 4. Response surface (a) and contour plots (b) for NH$_3$–N removal efficiency as a function of ozone dosage, (30 g/m^3), COD concentration, (250 mg/l) and reaction time, (60) min

Optimization of Semi-Aerobic Landfill Leachate Treatment Using Ozone 241

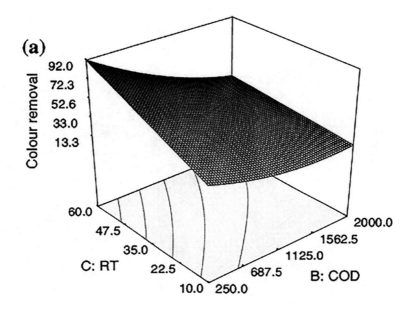

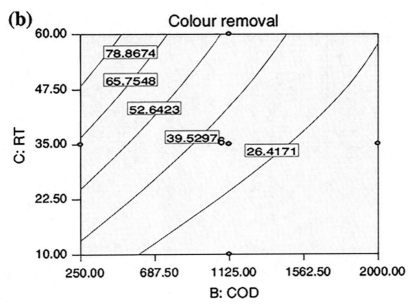

Figure 5. Response surface (a) and contour plots (b) for color removal efficiency as a function of ozone dosage, (30 g/m3), COD concentration, (250 mg/l) and reaction time, (60) min

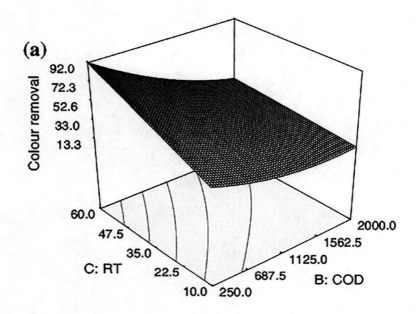

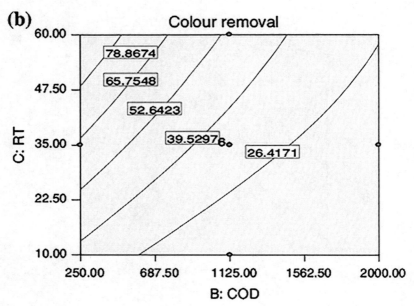

Figure 6. Response surface (a) and contour plots (b) for OC as a function of ozone dosage, (30 g/m³), COD concentration, (250 mg/l) and reaction time, (60) min

Optimization of Semi-Aerobic Landfill Leachate Treatment Using Ozone   243

Table 6. Optimization results for COD, NH_3–N Color maximum removal efficiency and minimum amount of OC.

NO	Ozone g/m³	COD mg/l	RT min	COD removal (%)	Ammonia removal (%)	Colour removal (%)	OC (kgO₃/kg COD)	Desirability
1	70	250	60	26.7	7.1	92	9.42	0.823
Lab. experiment			24.7	6.4	90.8	9.5		

11.4 CONCLUSION

Using ozone, the optimization of semi-aerobic stabilized landfill leachate treatment was investigated. The interaction between operational variables for the treatment optimization process, such as ozone dosage, COD concentration, and reaction time, was applied using RSM with CCD. Statistical analysis for the interaction of models' responses (COD, NH_3–N, color removal, and OC) was significant at P value less than 0.05. The optimum operational condition obtained 80 g/m³ of ozone gas applied on leachate with 250 mg/l COD concentration during 60 min reaction time was required to achieve 26.7, 7.1, and 92 % removal for COD, NH_3–N, and color, respectively. OC value (9.40 kgO₃/kg COD) was also obtained.

REFERENCES

1. Abu Amr SS, Aziz HA (2012) New treatment of stabilized leachate by ozone/ Fenton in the advanced oxidation process. Waste Manag. doi:10.1016/j.wasman. 2012.04.009
2. American Public Health Association (APHA) (2005) Standard methods for the examination of water and wastewater, 21th edn. American Public Health Association, Washington, DC
3. Amonkrane A, Comel C, Veron J (1997) Landfill leachates pretreatment by coagulation–flocculation. Water Res 31(11):2775–2782

4. Aziz HA, Alias S, Adlan MN, Asaari FAH, Zahari MS (2007) Colour removal from landfill leachate by coagulation and flocculation processes. Bioresour Technol 98:218–220
5. Aziz HA, Daud Z, Adlan MN, Hung YT (2009) The use of polyaluminium chloride for removing colour, COD and ammonia from semi-aerobic leachate. Int J Environ Eng 1:20–35
6. Baig S, Liechti PA (2001) Ozone treatment for biorefractory COD removal. Water Sci Technol 43:197–204
7. Bashir JKM, Hamidi AA, Yusoff MS (2011) New sequential treatment for mature landfill leachate by cationic/anionic and anionic/cationic processes: optimization and comparative study. J Hazard Mater 186:92–102
8. Beaman MS, Lambert SD, Graham NJD, Anderson R (1998) Role of ozone and recirculation in the stabilization of landfill leachates. Ozone Sci Eng 20(2):121–132
9. Chaturapruek A, Visvanathan C, Ahn KH (2005) Ozonation of membrane bioreactor effluent for landfill leachate treatment. Environ Technol 26:65–73
10. Christensen TH, Kjeldsen P, Bjerg PL, Jensen DL, Christensen JB, Baum A, Albrechtsen H, Heron G (2001) Biogeochemistry of landfill leachate plumes. Appl Geochem 16:659–718
11. Cortez S, Teixeira P, Oliveira R (2011) Manuel Mota Mature landfill leachate treatment by denitrification and ozonation. Process Biochem 46:148–153
12. Foul AA, Aziz HA, Isa MH, Hung YT (2009) Primary treatment of anaerobic landfill leachate using activated carbon and limestone: batch and column studies. Int J Environ Waste Manag 4:282–298
13. Geenens D, Bixio B, Thoeye C (2001) Combined ozone-activated sludge treatment of landfill leachate. Water Sci Technol 44:359–365
14. Geissen SU (2005) Experience with landfill leachate treatment in Germany. In: Workshop on landfill leachate: state of the art and new opportunities, INRST, Borj Cedria, Tunis, Tunisia
15. Goi A, Veressinina Y, Trapido M (2009) Combination of ozonation and the Fenton processes for landfill leachate treatment: evaluation of treatment efficiency. Ozone Sci Eng 31:28–36
16. Haapea P, Korhonen S, Tuhkanen T (2002) Treatment of industrial landfill leachates by chemical and biological methods: ozonation, ozonation hydrogen peroxide, hydrogen peroxide and biological post-treatment for ozonated water. Ozone Sci Eng 24:369–378
17. Hagman M, Heander E, Jansen JLC (2008) Advanced oxidation of refractory organics in leachate—potential methods and evaluation of biodegradability of the remaining substrate. Environ Technol 29:941–946
18. Ho S, Boyle WC, Ham RK (1974) Chemical treatment of leachate from sanitary landfills. J Water Pollut Control Fed 46:1776–1791
19. Huang SS, Diyamandoglu V, Fillos J (1993) Ozonation of leachates from aged domestic landfills. Ozone Sci Eng 15:433–444
20. Joglekar AM, May AT (1987) Product excellence through design of experiments. Cereal Foods World 32:857–868

21. Nordin MY, Venkatesh VC, Sharif S, Elting S, Abdullah A (2004) Application of response surface methodology in describing the performance of coated carbide tools when turning AISI 104 steel. J Mater Process Technol 145:46–58
22. Park J, Jin YS (2005) Effect of ozone treatment on ammonia removal of activated carbons. J Colloid Interface Sci 286:417–419
23. Read AD, Hudgins M, Harper S, Phillips JM (2001) The successful demonstration of aerobic landfilling: the potential for a more sustainable solid waste management approach. Resour Conserv Recycl 32:115–146
24. Renou S, Givaudan JG, Poulain S, Dirassouyan F, Moulin P (2008) Landfill leachate treatment: review and opportunity. J Hazard Mater 150:468–493
25. Rice RG (1997) Applications of ozone for industrial wastewater treatment—a review. Ozone Sci Eng 18:477–515
26. Rivas FJ, Beltrán F, Gimeno O, Acedo B, Carvalho F (2003) Stabilized leachates: ozone-activated carbon treatment and kinetics. Water Res 37:4823–4834
27. Schrab GE, Brown KW, Donnelly KC (1993) Acute and genetic toxicity of municipal landfill leachate. Water Air Soil Pollut 69:99–112
28. Tengrui L, AL-Harbawi AF, Bo LM, Jun Z (2007) Characteristics of nitrogen removal from old landfill leachate by sequencing batch biofilm reactor. J Appl Sci 4(4):211–214
29. Tizaoui C, Bouselmi L, Mansouri L, Ghrabi A (2007) Landfill leachate treatment with ozone and ozone/hydrogen peroxide systems. J Hazard Mater 140:316–324
30. Trebouet D, Schlumpf JP, Jaouen P, Quemeneur F (2001) Stabilized landfill leachate treatment by combined physicochemical–nanofiltration processes. Water Res 35(12):2935–2942
31. Wang F, Smith DW, El-Din MG (2003) Oxidation of aged raw landfill leachate with O3 only and O3/H2O2 and molecular size distribution analysis. In: Proceedings of the 16th World Congress of the International Ozone Association, IOA, Las Vegas, USA, pp 1–21
32. Wu JJ, Wu CC, Ma HW, Chang CC (2004) Treatment of landfill leachate by ozone-based advanced oxidation processes. Chemosphere 54:997–1003
33. Zazouil MA, Yousefi Z (2008) Removal of heavy metals from solid wastes leachates coagulation- flocculation process. J Appl Sci 8(11):2142–2147

CHAPTER 12

Removal of COD, Ammoniacal Nitrogen and Colour from Stabilized Landfill Leachate by Anaerobic Organism

MOHAMAD ANUAR KAMARUDDIN, MOHD SUFFIAN YUSOFF, HAMIDI ABDUL AZIZ, AND NUR KHAIRIYAH BASRI

12.1 INTRODUCTION

Landfilling is the primary means of municipal solid waste disposal in many countries worldwide owing to its economic advantages and minimum technology being practiced. Contamination of surface and ground water through leachate; soil contamination through direct waste contact or leachate; air pollution through the burning of wastes and the uncontrolled release of methane by anaerobic decomposition of waste (Aziz et al. 2010) are some of the effects of landfilling activities. In addition, landfill leachate produced from landfill sites consists of contaminated pollutants that is very difficult to deal with (Umar et al. 2010). If not properly treated and safely disposed, leachate can migrate to soil and subsoils which might cause severe damage to eco-system of land and receiving water. Landfill

---

Biophilic Cities Are Sustainable, Resilient Cities. © 2013 by the authors; licensee MDPI, Basel, Switzerland. Sustainability 2013, 5(8), 3328-3345; doi:10.3390/su5083328. Creative Commons Attribution license (http://creativecommons.org/licenses/by/3.0/).

leachate contains pollutants that can be categorized into four groups [dissolved organic matter, inorganic macrocomponents, heavy metals, and xenobiotic organic compounds (Kjeldsen et al. 2002)] as consequence of the existing waste disposal practice. Organic content of leachate pollution is generally measured by chemical oxygen demand (COD) and biochemical oxygen demand (BOD$_5$). Leachate is also rich in ammonia, halogenated hydrocarbons suspended solid, significant concentration of heavy metals and inorganic salts (Renou et al. 2008; Umar et al. 2010; Bashir et al. 2009; Aziz et al. 2010). According to Umar et al. (2010) for stabilized leachate, the COD content generally ranges between 5,000 and 20,000 mg/L. The BOD$_5$/COD ratio provides a good estimate of the state of the leachate and this ration for young leachate is generally between 0.4 and 0.5 (Kurniawan et al. 2006). The presence of concentrated pollutants in leachate has become one of the major issues to the landfill operator in obeying strict regulations for the safe disposal of leachate. Normally, the existence of high levels of pollutants such as ammonia in landfill leachate over a long period of time leads to a motivated algal growth, decreased performance of biological treatment systems, accelerated eutrophication, promoted dissolved oxygen depletion, and increased toxicity of living organisms in water bodies (Aziz et al. 2010).

Various process applications such as aerobic and anaerobic biological degradation, chemical oxidation, chemical precipitation, coagulation–flocculation, activated carbon adsorption and membrane processes have been reported by various authors (Aziz et al. 2010; Halim et al. 2009; Aghamohammadi et al. 2006; Deng and Englehardt 2006). Anaerobic treatment is one of the treatments in biological process which is ideally fitted for the pre-treatment of high strength wastewaters that are typical of many industrial facilities for the removal of the bulk of effluent containing high concentrations of COD, BOD and heavy metals. In favour of anaerobic treatment, biodegradation is carried out by the microorganisms, which can degrade organic compounds to carbon dioxide and sludge under aerobic conditions and to biogas (a mixture comprising chiefly CO$_2$ and CH$_4$) under anaerobic conditions (Ghasimi et al. 2010). Effective microorganism (EM) technology identified as one of the biological treatment that comprises the most effective application for heavy metal removal in industrial wastewaters (Zhou et al. 2008). Historically, EM was first

Removal of COD, Ammoniacal Nitrogen and Colour 249

discovered by Teruo Higa of the University of the Ryukyus in Japan (Higa and Parr 1994). Initially, Higa and Parr investigated the microorganisms that were useful in crop farming as they were seen as alternatives to the use of pesticides and fertilizers (Zhou et al. 2008). However, as far as the authors concern, technology of EM or anaerobic organism is mostly being used in pig, cattle, dairy, and poultry farming for purification of the wastewater (Higa and Parr 1994; Joo et al. 2006). The main activity in the anaerobic process is the creation of an antioxidant environment by anaerobic organisms, which is deemed crucial to the enhancement of the solid–liquid separation in cleaning water practice. However, its usage and the effectiveness on industrial wastewater such as leachate is very far limited and documented. Therefore, this study was carried out to determine the feasibility of anaerobic organism inoculation in biological treatment of stabilized landfill leachate. The main purpose of the existing research is to investigate the effect of leachate parameters removal namely COD, NH_3–N and colour by anaerobic organism cultures. The efficiencies of COD, NH_3–N and colour removal were tested via different pH of leachate sample, dosage and contact time of anaerobic organism via laboratory experimental set up.

12.2 MATERIALS AND METHODS

12.2.1 ANAEROBIC ORGANISM PREPARATION

Anaerobic organism solution was procured from Seberang Perai Municipal Council (Malaysia). The preliminary experimental work for obtaining the optimum ratio of brown sugar/fruit waste/distilled water conducted towards the removal of COD, NH_3–N and colour. The livelihood of the anaerobic organisms was also tested at the longest possible of preliminary experiment so that the microbial activity in the leachate sample lasts longer. To prepare anaerobic organism activated solution (AOAS), a ratio of 1:3:10 which was: 1 part of molasses/brown sugar; 3 parts of fruit waste; and 10 parts of distilled water were used. The mixture was amalgamated in a Teflon container and mixed thoroughly. The AOAS was fermented for 12 weeks. The survival of the microorganisms was largely dependent by the temperature of the environment. Therefore, the prepared mixtures were kept under minimum temperature fluctuations without direct sun light exposure. The off-gassing process occurred during the 7 days of fermenting with the released of carbon dioxide (CO_2). The entire process of fermentation was strictly monitored. In addition, the cover of the tank was ensured tightly closed to avoid generation of mold. The pH value was observed at the final duration of fermentation. A pH value in the range of 3–4 indicated that the fermentation process has completed. Next, the anaerobic organisms were kept in an airtight container to keep it anaerobic prior to use. Fundamentally, the activated anaerobic organism suspension is a mixture of molasses (brown sugar) and anaerobic organism in free chlorinated water or rice rinse water which provides the minerals for the multiplication of the microorganisms (Mohamad Yatim et al. 2009). The activated anaerobic organisms were then fermented in an anaerobic environment for 7–10 days of incubation. Prior to the experimental works, each of the containers was rinsed with 10 % (v/v) HCL and then deionised with water to remove background effects of the medium.

12.2.2 IDENTIFICATION OF ANAEROBIC ORGANISM

Identification of anaerobic organisms was carried out according to the manual (Gerhardt 1994) and the procedure recommended by effective microorganism (EM) Research Organization (Okinawa, Japan). Theoretically, anaerobic organisms are rich with selected species of microorganisms, which can be commonly found in many ecosystems, including lactic acid bacteria (LAB), yeasts, actinomycetes photosynthetic bacteria, and other types of organisms. The analyses were done by diluting fermented materials in phosphate buffer (pH 7.0) at appropriate levels and then spread-plated onto an appropriate agar media and incubated at 25–30 °C at least 1 week prior to observing the morphological and physiological characteristics. Initial laboratory investigation indicated that microbial groups of lactic acid bacteria, yeast and total bacteria count of 1.6×10^7, 1.0×10^4 and 1.2×10^7 cfu/mL, respectively, exists in the solutions.

12.2.3 LEACHATE SAMPLING

The leachate used in the experiment was collected from Pulau Burung Landfill Site (PBLS), which is situated in the North West of Malaysia. PBLS started its operation in 1991 as a semi-aerobic system complying with Level II sanitary landfill standards by establishing a controlled tipping technique. The site receives 1,800 tonnes of domestic waste daily which origins from Penang Island and Perai district. Table 1 shows the characteristics of the solid wastes at PBLS. In 2001, the landfill was upgraded to a Level III sanitary landfill by employing controlled tipping with leachate recirculation (Aziz et al. 2010). The site has a natural marine clay liner with daily covers practices. Characteristically, PBLS has surpassed more than 20 years of operation which in the methanogenic phase, and the leachate produced is referred to as mature, "stabilized" leachate (Bashir et al. 2009). The dark-coloured liquid, with high concentrations of COD and ammonium, and a low BOD_5/COD ratio is another indicator of stabilized leachate. The leachate sample was collected from September 2010 until March 2011. The samples were transported to the laboratory, and stored in a cold room at 4 °C to minimize biological and chemical changes prior to the experimental use (Bashir et al. 2010). The sampling procedure and preservation of samples were done according to the Standard Methods for the Examination of Water and Wastewater (APHA 2005). pH of the leachate sample was measured on-site using portable pH meter (Hach, sens ion 1, USA). The characteristics of the raw leachate from the oxidation pond of PBLS are shown in Table 2.

12.2.4 EXPERIMENTAL PROCEDURE

The study was focused on the preliminary determination of the condition that provides the optimum condition of anaerobic organisms in treating leachate. Prior to the experimental works, each of the containers was rinsed with 10 % (v/v) HCL and then deionised with water to remove the

Table 1. Solid waste characteristics at PBLS.

Waste characteristics	Amount (%)
Food	40
Plastic	22
Paper	10.5
Metals	2.5
Glass	3.25
Textile	3.5
Others	18.25
Total	100

background effects of the medium. Three different aspects of the experimental procedure were as follows:

12.2.4.1 INFLUENCE OF PH

To investigate the influence of pH by anaerobic organism in the leachate sample, six containers of 20-cm length and 10-cm height with a total volume of 4,000 mL made of Plexi glass were used. The containers were filled with 1,500 mL raw leachate sample and pH was adjusted from 5 to 9 for each of the containers by either 0.1 N hydrochloric acid (HCl) or 0.1 N sodium hydroxide (NaOH). Then, anaerobic organisms were added to leachate at ratio of 1:10 (150 mL of anaerobic organism) using measuring cylinder. The solution was stirred with a glass rod thoroughly to avoid chemical reaction with the leachate sample and the anaerobic organism activities in the container. A net grating was installed on the top of each container to allow anaerobic activity with the surrounding environment throughout the experimental period. The containers were kept for 24 days to ensure complete reaction between leachate sample and anaerobic organisms and the percentage removal of COD, NH_3–N and colour were eventually measured.

Removal of COD, Ammoniacal Nitrogen and Colour

Table 2. Raw leachate characteristics.

Parameters	1	2	3	4	5	6	Average	DOE standard
pH	7.81	7.67	7.34	7.65	7.12	7.51	7.52	5.5–9.0
Ammoniacal nitrogen (mg/L)	1,335	1,374	1,554	1,207	1,274	1,187	1,322	5
BOD_5 (mg/L)	302	287	269	280	242	273	276	20
COD (mg/L)	1,420	1,374	1,271	1,429	1,238	1,457	1,365	400
BOD_5/COD	0.21	0.21	0.21	0.20	0.20	0.19	0.20	0.05
Colour (ADMI)	1,568	1,547	1,761	1,644	1,543	1,455	1,586	200
Zinc (mg/L)	3.0	2.5	2.6	3.5	2.9	2.8	2.9	2.0
Iron (mg/L)	5.9	5.2	6.9	5.1	4.9	4.3	5.3	5.0
SS (mg/L)	258	237	249	298	301	168	250	100

12.2.4.2 INFLUENCE OF DOSAGE OF ANAEROBIC ORGANISM

Different dosage of anaerobic organisms at optimum pH was mixed in five containers each with the exact volume of 1,500 mL leachate sample. Different dosage ratio of 1:25, 1:20, 1:15, 1:10, and 1:5 which were 60, 75, 100, 150, and 300 mL of anaerobic organisms were filled in five containers, respectively, and kept under ambient air for 24 days. A net grating was installed on top of each container to allow anaerobic activity with the surround environment. The containers were kept in normal air condition in the range of 25–30 °C and the percent removal of leachate parameters was measured for COD, NH_3–N and colour, respectively.

12.2.4.3 INFLUENCE OF CONTACT TIME

To investigate the effect of removal efficiency of COD, by organic organisms, fixed amount of anaerobic organisms at optimum pH were prepared in 24 containers. A net grating was installed on top of each container to allow anaerobic activity with the surround environment. The sample was kept under normal air condition from day 1 to day 24 regimes. At the elapse of each set contact time, the treated leachate was filtered and analyzed for COD, NH_3–N and colour, respectively.

12.2.4.4 ANALYTICAL METHOD

Treated leachate samples were first filtered by using 0.45µ GC-50 glass micro fibre filters (Advantec, Japan) to retain fines particles from passing through. COD was measured using Calorimetric Method (5220-D) whereby the concentration of NH_3–N was measured based on the Nesslerization Method (Method: 8038) using UV spectrophotometer (HACH DR 2500, USA). Colour measurements were described as true colour, filtered using 0.45µ GC-50 (Advantec, Japan) assayed at 455 nm using DR 2500 HACH spectrophotometer. Method No. 2120C reports colour

Removal of COD, Ammoniacal Nitrogen and Colour

in platinum–cobalt (Pt–Co), the unit of colour being produced by 1 mg platinum/L in the form of the chloroplatinate ion (Al-Hamadani et al. 2011).

12.3 RESULTS AND DISCUSSIONS

12.3.1 RAW LEACHATE CHARACTERISTICS

Table 1 illustrates some of the leachate characteristics at PBLS. Even though the samples were collected on different days or at different times, the leachate characteristics demonstrate similar fashion and were largely comparable for most of the parameters. The average pH value of the leachate samples is 7.52. According to Umar et al. (2010), the pH of young leachate was less than 6.5, while for old landfill, the leachate has pH higher than 7.5. Initial low pH was due to high concentration of volatile fatty acids (VFAs). Similarly, stabilized leachate shows fairly constant pH with little variations and it may range between 7.5 and 9. The presence of significant amount of NH_3–N (1,322 mg/L) in the leachate indicates degradation of soluble nitrogen due to the decomposed waste. As a result, the concentration of NH_3–N increases with the increase in age of the landfill which was due to hydrolysis and fermentation of nitrogenous fractions of biodegradable refuse substrate (Umar et al. 2010). Higher concentration of NH_3–N enhances algae growth and promotes eutrophication due to decrease in dissolved oxygen content. Further, nitrification also leads to motivated algal growth, decreased performance of biological treatment systems, accelerated eutrophication, promoted dissolved oxygen depletion and increased toxicity of living organisms in water bodies (Aziz et al. 2010). According to Kurniawan et al. (2006), ammoniacal nitrogen was ranked as a major toxicant to living organisms, as established by various toxicity analyses using bioassays such as Salmo gairdnieri and Oncorhynchus nerka. The average values for BOD_5 and COD were 276 and 1,365 mg/L, respectively. Meanwhile, the calculated BOD_5/COD ratio was in the ratio of 0:20. In general, The BOD/COD

256   Sewage and Landfill Leachate

ratio indicates the changes in the amount of biodegradable compounds in the leachate (Warith 2002). Based on the ratio, it shows that the leachate was poorly biodegradable. The low BOD/COD ratio (0:20) also shows that the leachate was stable and difficult to be further degraded biologically (Foul et al. 2009). The high value of pH and low concentrations of COD, BOD_5 and heavy metals indicate that the leachate was in the phase of methane fermentation and classified as anaerobic phase (Salem et al. 2008). Normally, at the early stage of landfill operation, the presence of heavy metals are significant because of higher metal solubility whereby low pH caused by production of organic acids (Kulikowska and Klimiuk 2008). The colour concentration was reported as 1,586 mg/L, which was due to the low biodegradability of dissolved organic constituents in the leachate. Slow process of decomposition of buried waste has induced the percolation of liquid with the refuse. Therefore, dark-coloured liquid is an indicator of saturated organic content in the leachate. Overall, the leachate characteristics did not meet the permissible limit declared under Environmental Quality (Control of Pollution From Solid Waste Transfer Station and Landfill) Regulations 2009 (PU(A) 433) Second Schedule (Regulation 13) from Department of Environment, Malaysia.

12.3.2 OPTIMUM PH

Since the large majority of bacteria present growth optima at or near neutral pH values, most laboratory-based biodegradation studies have been carried out in this pH range. Figure 1 shows the variation of pH adjustment of leachate sample to the leachate parameters removal. From the figure, it was shown that anaerobic organisms played a significant role in removal efficiency of leachate parameters. By lowering the pH of leachate sample, COD percentage removal was increased from 39.7 to 43.7 %. Under acidic condition, all organic compounds were fully oxidized to carbon dioxide. In this case, although anaerobic organisms are not an oxidizing agent, microorganisms in the solution dissolved thoroughly in leachate sample that are responsible for the degradation of the organic content. Furthermore, the reduction of COD was due to the organic strength of the leachate which is affected by the reduction of the methanogenic bacteria. Increasing pH from 8 to 9 would lead to a slight increment of the latter from 49 to 51 %. Lowering pH of leachate sample reduced the amount of dissolved oxygen needed by aerobic biological organisms in leachate sample to break down the organic material present, which was known as the worse condition of anaerobic organisms' survival. In contrast, normal pH of leachate leads to higher removal of COD which comprises of 76.8 %. It was assumed that anaerobic organisms would tolerate in the normal pH of leachate. Similar finding was observed for the case of NH_3–N removal. Under acidic condition, NH_3–N removal was observed below 29 %, whereas slight increment removal was recorded when pH increased from 8 to 9. According to Rashid and West (2007), most microorganisms exhibit optimal growth at pH values between 6 and 8 and most could not tolerate at pH levels above 9 or below 4. The phenomenon of free ammonia inhibition of ammonia oxidizing bacteria (AOB) and nitrite oxidizing bacteria (NOB) also were found affected by pH adjustment. High pH would increase the free ammonia concentration and inhibits nitrifying activity, especially when the NH_4^+–N level was high (Zhang et al. 2007). Moreover, increasing the pH of leachate attributed to the higher concentration of NH_3–N generated from biodegradation process due to parts of protein were biodegraded and some nitrogen species were transferred to the form of NH_3 after bioreaction. These results suggest that the bacteria produce enzymes whose optimal activity occurs in different pH values. For the colour removal, the highest reduction was achieved when the leachate pH at 7 (46.2 %). As the pH of the leachate increased to alkaline state, further reduction was observed. This was due to condition of the organisms in the mixture of leachate which is susceptible when the pH of the leachate was normal. Since the present work intended to investigate the ability of anaerobic organisms to remove pollutants in a stationary condition of leachate sample, the excess negatively charged hydroxide ions have to compete with inorganic constituents available in the leachate. Therefore, colour removal was not favourable in this condition. This result was in agreement with the work reported by Oliveira et al. (2009).

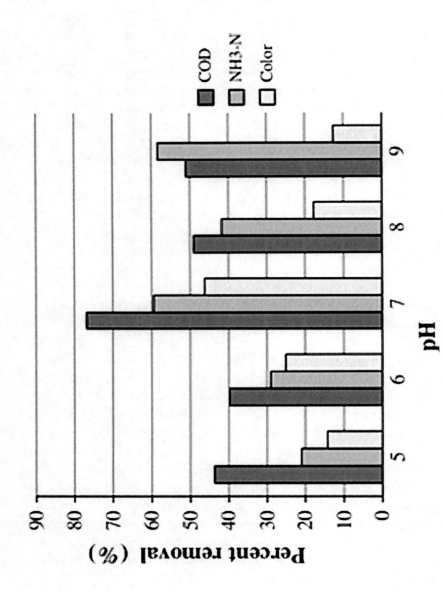

Figure 1. Influence of pH for the removal of COD, NH_3–N and colour.

12.3.3 OPTIMUM DOSAGE

Figure 2 shows the influence of anaerobic dosage to the removal of leachate parameters. Briefly, removal of COD and NH_3–N were directly proportional to the anaerobic organism dosage to a certain level. The presence of anaerobic organism in the leachate sample at ratio 1:15 (100 mL) could remove up to 42 % of the COD concentration followed by dosage of 75 mL (1:20) and 100 mL (1:15), respectively. Meanwhile, the anaerobic organism dosage of 150 mL (1:10) and 300 mL (1:5) could only remove COD concentration at 17.6 and 4.2 %, respectively. Therefore, based on the observation from the figure, the highest removal of COD concentration achieved when 100 mL of anaerobic organism was introduced in the leachate sample. Generally, as the dosage of anaerobic organism increased, the removal of COD concentration was increased due to the increasing of active bacteria and organic acid in anaerobic organism including *Bacillus subtilis* which decomposes and digests the organic residuals in leachate. Meanwhile, the lactic acid bacteria increase the carbohydrate metabolism by producing lactic acid and other antimicrobial products. Those antimicrobial products having antibacterial properties that inhibited the growth of pathogens and other non-beneficial microorganisms during the biodegradation of organic particle in leachate sample. From the figure, the removal efficiency of NH_3–N was reduced when incrementing the amount of anaerobic organisms was introduced. At dosage of 100 mL, highest removal of NH_3–N was achieved at 57.21 %. In contrast, reducing dosage of anaerobic organisms resulted low removal of NH_3–N. Low removal in NH_3–N concentration leads to motivated algal growth, accelerated eutrophication and rendered the livelihood of the microorganisms without food availability. These reductions could be attributed to the microbial utilization of these nutrients. Inorganic nitrogen variation should be similar to COD, since COD is consumed to oxidize the inorganic chemicals. The removal pattern for colour was observed similar to COD and NH_3–N, respectively. The highest colour reduction was observed when the anaerobic organism dosage was 100 mL (52 %). As the dosage of the anaerobic organism increases,

260   Sewage and Landfill Leachate

lower removal of the colour was observed at dosage 1:5. Initially, the culture mixtures available were adequate during the microbial activity in the leachate mixture. Consequently, addition of anaerobic organisms was found to attribute mainly towards the stabilized biological activity of the leachate sample (Zouboulis et al. 2001). This would hinder the breakdown process of dissolved organic constituents which deplete the microbial activity in the leachate that leads to formation of dark liquid in the container.

12.3.4 OPTIMUM CONTACT TIME

The removal efficiency of COD and NH_3–N is shown in Fig. 3. At initial interval of treatment, slight removal was observed for COD and NH_3–N where only 1.5 and 8.1 % reductions were recorded. The data show that the treatment process provides a consistent high efficiency of COD removal until day 12 accounted for 42 %. Similar observation was also recorded in the case of NH_3–N where the removal percentage was increased from 1.5 to 42.1 % at day 14. Despite the fluctuations of COD and NH_3–N concentrations on the first 14 days of treatment, the removal pattern shows a gradual reduction for the case of COD. Hence, a slow rate of biodegradation might have caused the COD increase over the time. From the figure, a slow dropped off percent removal of NH_3–N occurred at day 18. Increasing the contact time of anaerobic organisms allowed the microorganism in the leachate sample mixed assisted by acclimatize with the indigenous microbial population in the leachate. As the period of leachate sample prolonged, COD and NH_3–N removal was observed reduced significantly. The differences in reducing abilities of each leachate may be attributed to differences in bacteria species, bacterial biomass concentration or organic type, as well as landfill conditions such as waste content and landfill age. Similarly, microorganisms are often sensitive to their environment. Likewise, high concentration of NH_3–N increased nitrification reduced the rate of biodegradation of leachate. According to Umar et al. (2010), higher concentrations of ammonia are also known to enhance algal growth, promote eutrophication due to

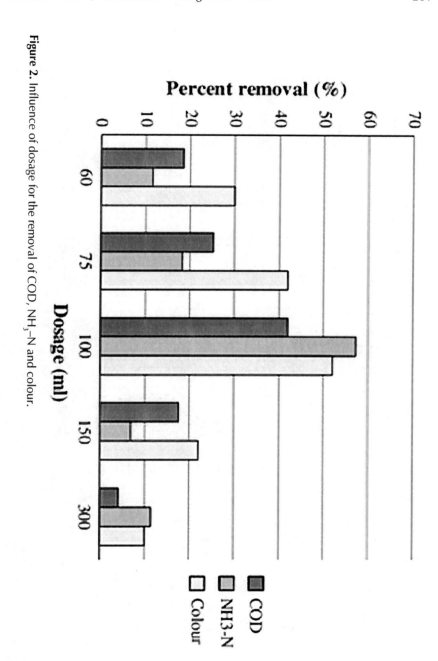

Figure 2. Influence of dosage for the removal of COD, NH_3–N and colour.

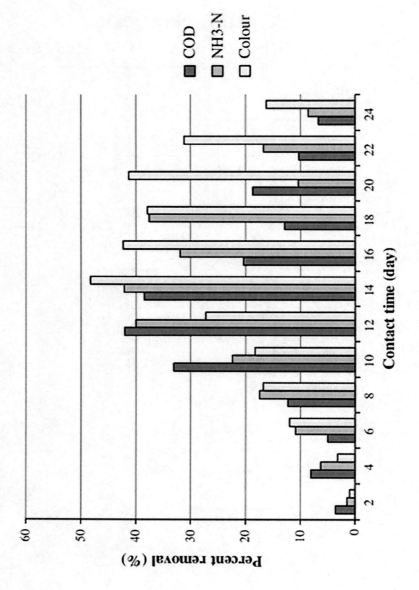

Figure 3. Influence of contact time for the removal of COD, NH_3-N and colour.

decreased dissolved oxygen. Hence, microbial activity was further depleted over time. In addition, longer contact time of anaerobic organisms and leachate sample will reduce the ability of different microorganism to react efficiently with the leachate pollutants, since more oxygen depleted in the containers. The mechanism involved suggests that as anaerobic organism was a mixed culture of many species of microorganisms, some of which can transform from NH_4^+ to NO_3^-, thereby decreasing the potential for N-fraction (Javaid 2010). Meanwhile, it was observed that the highest colour removal was recorded at 14 days (48.3 %). Although the reduction of colour was fluctuated after 14 days, the anaerobic organisms continue to decolorize until the 24 days of experiment at slow rate. The possible justification of this behaviour was governed by the rejuvenation of microbial activity in the leachate sample. When the period of incubation increases, longer duration of active organisms in the leachate sample with the surround environment which promotes the active cell of microorganism activity entirely.

12.3.5 OPTIMUM VARIABLE CONDITION

In assessing the optimum variable conditions of the anaerobic organisms, the evaluation of the preliminary experiment was prerequisite. Since the variable conditions produce different removal efficiency, the selection of the variables was based on the highest removal of COD, NH_3–N and colour. Table 3 shows the optimum condition of anaerobic organism preparation. It can be observed that the optimum condition of the anaerobic organism achieved at pH 7, dosage of 100 mL and required 14 days to reduce the COD, NH_3–N and colour pollutants to 65.5, 60.2 and 46.3 %, respectively. The final concentrations of the pollutants are: COD, 470 mg/L; NH_3–N, 526 mg/L; colour, 851 ADMI. We can observe that the final concentration of the treated leachate sample still exceeded the permissible discharge limit regulated by Malaysia DOE (COD, 400 mg/L; NH_3–N, 5 mg/L; colour, 200 ADMI).

Table 3. Optimum variable condition of anaerobic organisms.

Optimum condition			Percent removal		
pH	Dosage	Contact time (day)	COD	NH_3–N	Colour
7	100 mL	14	65.5	60.2	46.3

12.4 CONCLUSIONS

In this study, biological treatment of leachate by anaerobic organisms under ambient condition was investigated. The biological treatment process was found to be feasible for treating the leachate from stabilized landfill. COD, NH_3–N and colour-removal efficiencies under neutral pH of leachate sample were found in the range of 76.8, 54.4 and 46.2 %, respectively. Increasing the dosage of anaerobic organisms led to a higher removal of COD from 18.5 to 42 %, respectively. In addition, when equal dosage was tested for NH_3–N, up to 57 % removal was recorded. As for the colour removal, similar pattern was observed for the removal efficiency. In this case, 100 mL dosage of anaerobic organism was found optimum with 52 % colour reduction. However, further increasing of anaerobic organism resulted in drastic reduction of these parameters between 4.2 and 11.4 %, respectively, for COD, NH_3–N and colour. Meanwhile, removal pattern of COD, NH_3–N and colour increased from day 2 and reached maximum at day 14. However, fluctuation pattern was observed starting at day 16 until day 24 for COD and NH_3–N, respectively. Meanwhile, the optimum condition of the anaerobic organism achieved at pH 7, dosage of 100 mL and required 14 days to reduced COD, NH_3–N and colour pollutants to 65.5, 60.2 and 46.3 %, respectively. It also indicates that the actual contact time was reduced from 24 to 14 days when the replication of the experiment was conducted, which proves that anaerobic organism could shorten the process of microbial activity in leachate sample. As a result, the application of anaerobic organisms was effective for biological treatment of landfill leachate that simultaneously reduces high concentration of pollutants from landfill. However, the optimal conditions of AOAS were not able to reduce the concentration of leachate sample to a stipulated permissible discharge limit enforced by Malaysia DOE. The partial treatment system such as aeration process can be employed to improve oxidation state of organic content in the leachate sample provided that ammonia stripping is designated after the AOAS treatment to assist denitrification of ammonia content. Since AOAS is a biological process, the application of this treatment can be designated in the pre-treatment of leachate. In this case, the AOAS treatment can be conducted in a rector supplied with sprinkler system. At a certain time of period, the AOAS will be emitted in the reactor by the sprinkler system and the microbial activity occurs when the aerated leachate flows into the reactor. Conclusively, the AOAS is considered as partial treatment system for leachate remediation process considering its ability to reduce leachate contents. However, more investigation should be conducted to explore the ability of AOAS in full-scale reactor system so that the ability of this microbial system can be optimized.

REFERENCES

1. Aghamohammadi N, Aziz HA, Isa MH (2006) Removal of iron from semi-aerobic landfill leachate by activated Sludge-activated carbon process. UTP Publication http://eprints.utp.edu.my/1492/1/ Assessed 10 May 2012
2. Al-Hamadani YAJ, Yusoff MS, Umar M, Bashir MJK, Adlan MN (2011) Application of psyllium husk as coagulant and coagulant aid in semi-aerobic landfill leachate treatment. J Hazard Mater 190:582–587
3. APHA (2005) Standard method for the examination of water and wastewater, 21st edn. American Public Health Association, Washington
4. Aziz SQ, Aziz HA, Yusoff MS, Bashir MJK, Umar M (2010) Leachate characterization in semi-aerobic and anaerobic sanitary landfills: a comparative study. J Environ Manage 91:2608–2614
5. Bashir MJK, Isa MH, Kutty SRM, Awang ZB, Aziz HA, Mohajeri S, Farooqi IH (2009) Landfill leachate treatment by electrochemical oxidation. Waste Manage 29:2534–2541
6. Bashir MJK, Aziz HA, Yusoff MS, Aziz SQ, Mohajeri S (2010) Stabilized sanitary landfill leachate treatment using anionic resin: treatment optimization by response surface methodology. J Hazard Mater 182:115–122

7. Deng Y, Englehardt JD (2006) Treatment of landfill leachate by the Fenton process. Water Res 40:3683–3694
8. Foul AA, Aziz HA, Isa MH, Hung YT (2009) Primary treatment of anaerobic landfill leachate using activated carbon and limestone: batch and column studies. Int J Environ Waste Manag 4:282–298
9. Gerhardt P (1994) Methods for general and molecular bacteriology. American Society for Microbiology, Washington
10. Ghasimi SMD, Idris A, Chuah TG, Tey BT (2010) Semi-continuous anaerobic treatment of fresh leachate from municipal solid waste transfer station. Afr J Biotechnol 8:2763–2773
11. Halim AA, Aziz HA, Megat Johari MA, Ariffin KS (2009) Removal of ammoniacal nitrogen and COD from semi-aerobic landfill leachate using low-cost activated carbon zeolite composite adsorbent. Int J Environ Waste Manag 4:399–411
12. Higa T, Parr JF (1994) Beneficial and effective microorganisms for a sustainable agriculture and environment. International Nature Farming Research Centre Atami, Japan. http://www.emturkey.com.tr/TR/dosya/1-314/h/54-parrhigabkltcf12 0on20em.pdf. Assessed 15 May 2012
13. Javaid A (2010) Beneficial microorganisms for sustainable agriculture. Genetic Eng Biofertil Soil Quality Organic Farm. doi:10.1007/978-90-481-8741-6_12
14. Joo HS, Hirai M, Shoda M (2006) Piggery wastewater treatment using Alcaligenes faecalis strain No. 4 with heterotrophic nitrification and aerobic denitrification. Water Res 40:3029–3036
15. Kjeldsen P, Barlaz MA, Rooker AP, Baun A, Ledin A, Christensen TH (2002) Present and long-term composition of MSW landfill leachate: a review. Crit Rev Env Sci Tec 32:297–336
16. Kulikowska D, Klimiuk E (2008) The effect of landfill age on municipal leachate composition. Bioresource Technol 99:5981–5985
17. Kurniawan TA, Lo W, Chan G (2006) Physico-chemical treatments for removal of recalcitrant contaminants from landfill leachate. J Hazard Mater 129:80–100
18. Mohamad Yatim J, Rahman WA, Aizan W, Sam M, Rahman A (2009) Characterisation and effects of the effective micro-organics (EM) and industrial waste (IW) materials as a partial mixture in concrete.UTM Press, Skudai. http://www.eprints. utm.my/9710/. Accessed 12 May 2012
19. Oliveira PL, Duarte MCT, Ponezi AN, Durrant LR (2009) Use of Bacillus pumilus CBMAI 0008 and Paenibacillus sp. CBMAI 868 for colour removal from paper mill effluent. Braz J Microbiol 40:354–357
20. Rashid M, West J (2007) Dairy wastewater treatment with effective microorganisms and duckweed for pollutants and pathogen control. Nato Sec. doi:10.1007/ 978-1-4020-6027-4_10
21. Renou S, Givaudan JG, Poulain S, Dirassouyan F, Moulin P (2008) Landfill leachate treatment: review and opportunity. J Hazard Mater 150:468–493
22. Salem Z, Hamouri K, Djemaa R, Allia K (2008) Evaluation of landfill leachate pollution and treatment. Desalination 220:108–114
23. Umar M, Aziz HA, Yusoff MS (2010) Variability of parameters involved in leachate pollution index and determination of LPI from four landfills in Malaysia. Int J Chem Eng. doi:10.1155/2010/747953

24. Warith M (2002) Bioreactor landfills: experimental and field results. Waste Manage 22:7–17
25. Zhang S, Peng YZ, Wang SY, Zheng SW, Guo J (2007) Organic matter and concentrated nitrogen removal by shortcut nitrification and denitrification from mature municipal landfill leachate. J Environ Sci 19:647–651
26. Zhou S, Wei C, Liao C, Wu H (2008) Damage to DNA of effective microorganisms by heavy metals: impact on wastewater treatment. J Environ Sci 20:1514–1518
27. Zouboulis A, Loukidou M, Christodoulou K (2001) Enzymatic treatment of sanitary landfill leachate. Chemosphere 44:1103–1108

AUTHOR NOTES

CHAPTER 1

Acknowledgments
Authors would like to acknowledge Universiti Sains Malaysia for the financial support provided under RUI-USM 1001/PAWAM/814166 and Ministry of Higher Education for the scholarship awarded under MyBrain15.

CHAPTER 2

Acknowledgments
The authors wish to thank the Poiana Waterworks Ltd. for technical and economic support and the Regione Friuli Venezia Giulia administration; a particular thank goes to Irene Duse and Andrea Fattori for their laboratorial support.

Author Contributions
Claudia Bruna Rizzardini has developed this work in the framework of her Ph.D. Thesis. Daniele Goi, as the scientific expert, has supervised the overall work.

Conflicts of Interest
The authors declare no conflict of interest.

CHAPTER 3

Acknowledgments
The present work was part of the project named "Manejo integral de residuos líquidos y sólidos generados por la limpieza de tanques sépticos, sanitarios portátiles y aguas residuales de la nixtamalización" sponsored

by Fondos Mixtos Gobierno del Estado de Yucatán–Consejo Nacional de Ciencia y Tecnología.

Conflicts of Interest

The authors declare no conflict of interest.

CHAPTER 4

Acknowledgments

The authors want to thank University of Davis (CA, USA) for experimental support. Also, Marta Otero acknowledges financial support from the Spanish Ministry of Science and Innovation (RYC-2010-05634).

CHAPTER 5

Acknowledgments

Financial support was provided by the Polish Ministry of Science and Higher Education Grant No.: N305 320 636 in the years 2009–2011 and N305 327 439 in years 2010–2011 and by the European Union within the European Social Fund in project "InnoDoktorant—Scholarships for PhD students, 2nd edition". Help of Ms. Iwona Henke in performing strain isolation is greatly acknowledged. We would like to thank Daniel Szopiński for the artwork. In addition, we would like to thank Aleksandra Markiewicz for her assistance with bioinformatics.

CHAPTER 6

Acknowledgments

The Authors wish to thank the Fondazione Trentina per la Ricerca sui Tumori, and especially to the De Luca family for the support to this research. The Authors are also grateful to the Environmental Protection Agency of the Province of Trento (APPA), and especially, Mr. Maurizio Tava for his precious contribution. Special thanks to Dr. Paola Foladori for her important support in the premises of this activity, Ms. Roberta Villa and

Author Notes 271

Mr. Alessandro Chistè for their assistance in the sampling campaign, Mr. Paolo Andreatta for the precious data about the wastewater treatment plants, Dr. Werner Tirler and his research group for the analysis of the samples.

CHAPTER 7

Acknowledgments
The authors of this paper gratefully acknowledge the Swedish Energy Agency for financial support (STEM P30686-1). Hanna Söderström, Department of Chemistry, Umeå University, is thanked for help regarding the sampling of air.

Acknowledgments
The authors declare no conflict of interest.

CHAPTER 8

Conflicts of Interest
The authors declare no conflict of interest.

CHAPTER 9

Acknowledgments
The authors thank financial support from AMGA Spa, Udine and Passavant Impianti Spa, Milan. They are also grateful to Dott. Stefano Turco and Mr. Aldo Bertoni for laboratory help.

CHAPTER 10

Acknowledgments
We sincerely thank the Department of Science and Technology, New Delhi for the financial support rendered to carry out the research work. We also thank the Corporation of Tiruchirappalli for the permission to carry out the studies at the open dump sites.

CHAPTER 11

Acknowledgments

We acknowledge the staff and team of technicians from the School of Civil Engineering for their valuable help in facilitating and supporting the current work.

INDEX

A

absorbable organic halogens (AOX) 37, 39

Actinomycetes 13, 250

adsorption 6, 15–16, 18, 93, 177, 248

advance oxidation process (AOP) 20–21, 174

aerobic x, xx, 5, 7–9, 12–13, 19, 52, 89, 225–228, 243, 248, 251, 257

air x, xvii–xviii, xx–xxi, 22–23, 52, 66–67, 69–70, 72–73, 75, 77–78, 83, 105, 108, 113, 116, 132, 134, 140–142, 150, 176, 247, 254, 271
  flotation 22
  pollution 116, 247

algal growth 248, 255, 259–260

alkali xvii, 69–72, 76

alkalinity xx, 202, 209

alumina-silicate 69

ammonia xv, xvii, 14, 23, 70–72, 150–151, 159–160, 162, 165, 169, 198, 226, 228, 231, 248, 257, 260, 265

ammonium oxidation 12, 151

anaerobic x, xvi, xx–xxi, 7–9, 12–15, 19, 33, 66, 151, 169, 205, 247–252, 254, 256–257, 259–260, 263–265

antibacterial agents 122

Ardern, Edward 32

Arthrobacter 93

ash 64, 66–67, 69, 72–73, 76–77, 122

asthma 211

atomic absorption spectrophotometer (AAS) 203

B

Bacillus subtilis 259

bacteria xvii–xviii, 9, 12, 15, 32, 56, 84–85, 89, 93–94, 124, 135, 250, 256–257, 259–260

bioaccumulation 104–105, 124, 137, 173

biochemical oxygen demand (BOD) x, xix, 152, 155, 161, 176, 203, 207, 227, 248, 255–256

biodegradation xvii–xviii, 7, 9, 14, 42, 83–86, 89, 94–96, 173, 248, 256–257, 259–260

biogas 248

biomass 9, 12, 64, 72, 83, 85, 98, 260

bioreactor(s) 9, 13–14, 150
  membrane 14, 150

Bureau of Indian Standards (BIS) xx, 208–209, 212–213, 221

Burkholderia 13

C

cadmium (Cd) 35–36, 40, 42, 122, 155, 203–204, 207, 212, 214, 221

California 65

cancer 104

carbon xix, 6, 13, 16, 18, 20, 22, 66, 88, 90, 122, 176, 203, 226, 231, 248, 250, 256
  dioxide (CO_2) 23, 69, 78, 151, 226, 248, 250, 256

carcinogenicity 104, 124

cerium 175–176, 180, 186

oxide (CeO_2) 176, 179, 182–183, 186–187
chemical oxygen demand (COD) x, xix–xx, 6–7, 9, 12–16, 18–23, 150, 152, 155–156, 158, 161, 165, 168–169, 176, 178, 183–185, 187, 203, 207, 226–228, 230–234, 236–243, 247–249, 251–252, 254–264
chloride xvii, 15, 20, 84–85, 89–90, 94, 97, 152, 202, 207, 209–211, 220
chlorine 15, 20, 152
chromium x, xix, 152, 155–156, 165, 198
coagulation (coagulant) 6, 15, 18–20, 22, 174, 248
cold gas 64, 76, 79
color removal 23, 175, 231, 233–234, 236–238, 241, 243, 249, 257, 263–264
composting xvi–xvii, 33, 49–50, 52–54, 56, 58–59
copper (Cu) 35–37, 40, 42, 155, 165, 177, 198, 203–204, 207, 212, 215, 221
cropland 31
Cyperus haspan 14

D

dairy 35, 104, 106, 249
Davyhulme Sewage Works 32
denitrification 12, 151, 265
Denmark 33
di-2-(ethyl-hexyl) phthalate (DEHP) 35, 37–38
dibenzo-p-dioxins and furans (PCDD/F) 35, 37, 64, 103–107, 110–111, 114
dichloromethane 141
diesel 175–176
diseases 50, 56, 200, 212

dissolved organic matter (DOM) 20, 248
DNA 88–89

E

earthworms 124
electro-oxidation xix, 149–152, 162, 165, 168–169
electrochemical treatments x, xix, 151–152
Empty Bed Contact Time (EBCT) 18
energy consumption 168
England 31
ethinyl estradiol (EE2) xviii, 122–125, 127, 132, 135, 137, 141
Europe 31–33
  Northern 33
  Western 31
European xvi, 32–33, 37, 39, 42–43, 104, 123, 270
  Commission 33, 39, 42
  Community 33, 123
  Member States 37
  Union (EU) xvi, 33, 39, 123, 137, 270
extractable organic halogen (EOX) 35, 37–38

F

faecal coliforms xvii, 50, 52–53, 56, 58–59
farmlands x, 105
fatty acids 93, 205, 207, 255
Fenton process x, 22, 173, 175–176, 188
fermentation 250, 255–256
fish 104, 124
flame retardants 122–123
Flavobacteriaceae 97

Index

flocculation 6, 16, 18, 22, 161, 165, 248

fluorisil 142

fluoroquinolone antibiotics (FQs) xviii, 122–125, 127, 132, 135, 138, 140

food x, xviii, 31, 50, 104–108, 110, 113–115, 200, 204, 259

forest x, 57–58, 121–122, 124, 127, 137–138, 142

fulvic acid (FA) 6–7, 23

G

gasification ix, xvii, 63–64, 66–67, 69, 72–75, 78–79

Germany 33, 56, 85, 228

glass 69, 108, 177, 200, 204, 252, 254

Greece 33

groundwater x, xv, xix–xxi, 197–203, 208, 210–219, 221

H

halogens 6, 37

health xvi–xvii, xx–xxi, 33–34, 37, 52, 103, 165, 173, 197–199, 208–209, 226, 228

heavy metals xv–xvi, xx, 6, 14, 32–33, 35–37, 39–40, 42–43, 64, 122, 197, 199, 202–204, 212, 220–221, 226, 248, 256

helminth eggs xvii, 52–53, 55–56, 58

high resolution mass spectrometry (HRMS) 108, 141–142

humic acid 16

humus xvi, xviii, 52, 124–125, 127, 132, 135–138, 140–142

hydrocarbons xviii, 37, 70, 94, 122, 127, 141, 198, 226, 248

hydrochloric acid (HCl) 250–252

hydrogen peroxide (H_2O_2) 152, 169, 174, 176, 178, 183–186, 231

hydroxide (OH) 20, 23, 142, 174, 185–186, 207, 252, 257

hygiene products 122

I

imidazolium 83–85, 95

industrial solid waste(s) 225

International Agency for the Research on Cancer (IARC) 104

ionic liquids (ILs) ix, xvii, 83–85, 93–94, 97

iron (Fe) xix, 16, 22–23, 152, 161, 165, 174–177–178, 182–184, 187–188, 198, 207, 212–213

oxide (Fe2O3) 182–183, 187

Italy (Italian) v, xvi, 33–34, 38, 42, 105, 153, 155, 158, 165, 176

J

Japan 249–250, 254

L

lead (Pb) xx, 23, 35–36, 40, 94, 97, 137, 155, 175, 198, 204, 203–204, 207, 212, 217, 221, 257

dioxide (PbO_2) 23

linear alkylbenzene sulfonates (LAS) 35, 37–38

lipopolysaccharide (LPS) 93

liquid chromatography 141

lung cancer 104

M

magnesium 20, 198, 202, 206, 208, 220

Malaysia 3–4, 7–8, 19, 227, 249, 251, 256, 263, 265, 269

manganese (Mn) 198, 203–204, 207, 212, 216, 221

metal(s) xv–xvi, xx, 6, 14–15, 20, 32–33, 35–37, 39–43, 64, 83, 104, 122, 138, 150, 152, 155–156, 162, 168, 174, 197–200, 202–204, 207, 212, 220–221, 226, 248, 256
- heavy xv–xvi, xx, 6, 14, 32–33, 35–37, 39–40, 42–43, 64, 122, 197, 199, 202–204, 212, 220–221, 226, 248, 256
- ions 152, 174
- noble 162
- toxic 36, 41

methanol 22, 70, 140–141

Mexico xvi, 49–51, 53

Microbacteriaceae 97

Microbacterium keratanolyticum 93–94

Micrococcaceae 97

Moraxellaceae 97

municipal solid waste (MSW) 3–4, 50, 149–150, 153, 200, 204, 206, 220, 247

mutagenicity 124

N

Netherlands 33

nitrification 12, 151, 255, 260

nitrogen xv, xx, 12–14, 64, 75, 121–122, 151–152, 159–160, 162, 168–169, 177, 198, 247, 255, 257, 259

Nocardiaceae 97

non-Hodgkin lymphoma 104

nonylphenol and nonylphenolethoxyl-ates (NPE) 35, 37

O

organic pollutants x, xviii, 6, 37, 39, 103, 137–138, 174

oxygen x, xix, 6, 9, 12, 15, 23, 52, 66, 72, 75, 78, 84, 89, 95, 150, 152, 175, 177–178, 180, 186–187, 198, 203, 226, 228, 248, 255, 257, 263

ozone x, xx, 20, 22, 174, 225–232, 238–243

P

pastures x, 105

peroxide 23, 152, 169, 174, 176, 184–186, 188, 231

Persistent Organic Pollutants (POPs) x, xviii, 6, 37–38, 103–107, 113–115, 138

persulfate (K2S2O8) 23

pharmaceutical residues 122

Piedmont 153

pine needles 134, 140–142

plasticizers 122

Poland 86, 88

poly-magnesium–aluminum sulfate (PMAS) 20

polyaromatic hydrocarbons (PAHs) xviii, 38–39, 122–125, 127, 137–138, 141, 198

polyaromatic hydrocarbons (PAHs) xviii, 38–39, 122–125, 127, 137–138, 141

polybrominated biphenyl ethers (PB-DEs) xviii, 122–124, 127, 132, 135, 137–138, 141–142

polybrominated diphenyl ethers (PB-DEs) xviii, 122–124, 127, 132, 135, 137–138, 141–142

polychlorinated biphenyls (PCBs) xviii, 6, 38–39, 103–104, 107–108, 110, 113, 115–116, 122–125, 127, 137–138, 141

polychlorinated dibenzo-p-dioxins (PCDDs) 38, 103

Index

polychlorinated dibenzofurans (PCDFs) 38, 103

polycyclic aromatic hydrocarbons (PAH) 35, 37, 124, 131–132, 135, 138, 142

polyelectrolytes 107–108

porines 93

potassium 20, 23, 198

poultry 249

precipitation 5, 7, 21, 149–150, 174, 202, 207, 248

Pseudomonas 13

pulpwood 121

R

rainfall 49, 199

rare earth 175

response surface methodology (RSM) x, xx, 18–19, 225–226, 230, 243

Rhodococcus 93

S

saline xx, 14, 86, 152, 208

salmonella xvii, 50, 52–53, 56, 58–59

sediment 135, 198

sequencing batch reactor (SBR) 12, 21, 23

sewage farms 31

sewer systems 5, 31

Shewanellacae 97

sludge ix–x, xvi–xix, 5, 12–13, 22–23, 31–43, 47, 49–50, 52–53, 55–59, 63–67, 70, 72–73, 76, 78–79, 83–86, 89, 95–98, 101, 103, 105–108, 110–116, 121–125, 132, 135, 137–138, 140–142, 150, 156, 175, 248

soap 208

sodium hydroxide (NaOH) 142, 252

soft-tissue sarcoma 104

soil(s) ix–x, xvi, xviii, 4–5, 20, 31–36, 39–43, 50, 52, 55, 57–59, 83–85,

104, 115, 121–122, 124–125, 127, 132–133, 135, 137–138, 140–142, 165, 198, 247

Spain 33

Sphingomonas paucimobilis 84

Standard Methods for the Examination of Water and Wastewater 53, 176, 227–228, 251

steel xviii, 23, 67, 69, 104–107, 109–110, 113–116, 154, 161–162, 165, 207

sulfate(s) xx, 15, 20, 202, 204, 212, 220

sulfuric acid (sulphuric acid) 70, 108, 142

Sweden 33, 121–123, 138

Swedish xviii, 121–123, 135, 142, 271

  Pharmaceutical Industry Association 123

  Society for Nature Conservation 123

T

Tamil Nadu x, xix, 199, 204

tank reactor 13

tar xvii, 64, 70–72, 76, 78–79

textiles 123, 200, 204

timber 121

titanium 162

toilets 31

toluene 141, 198

total organic carbon (TOC) xix, 6, 13, 16, 176, 178, 183–185, 187–188, 203, 212–213

tourist(s) 107–108

Trametes versicolor 14

triclosan (TCS) xviii, 122–125, 127–128, 132, 135, 137–138, 141–142

U

United States 31–32

U.S. Environmental Protection Agency 108, 123, 155

V

viruses 32
volatile organic acids 6

W

World Health Organization (WHO) ix, xx, 103, 110–113, 116, 208–209, 212–213, 221

X

xenobiotics xvii, 84–85, 94–95, 248

Y

Yucatán Península 49

Z

zinc (Zn) 35–36, 40, 42, 155, 203–204, 207, 212–213, 218, 221
zirconia xix, 176, 178, 180, 183, 186–188